ICE-CORE DRILLING

John F. Splettstoesser, Editor

Proceedings of a Symposium, University of Nebraska, Lincoln, 28-30 August 1974

University of Nebraska Press
Lincoln and London

"The Polar Ice-Core Storage Facility at USA CRREL" by Chester C. Langway, Jr. previously appeared in the *Antarctic Journal of the United States,* v. 9, no. 6, 1974. "General Considerations for Drill System Design" by Malcolm Mellor and Paul V. Sellmann was printed and distributed in June 1975 as Technical Report 264 of Cold Regions Research and Engineering Laboratory, Hanover, New Hampshire.

In the interest of timeliness and economy, this work was printed directly from camera-ready copy prepared by the editor.

Manufactured in the United States of America

PREFACE

In response to suggestions that an attempt be made to bring together the scientists and engineers from the several nations that are currently involved in ice and ice-core drilling in the polar regions, the Ross Ice Shelf Project Management Office at the University of Nebraska hosted an Ice-Core Drilling Symposium on August 28-30, 1974, at the Nebraska Center for Continuing Education, Lincoln, Nebraska. The emphasis of the Symposium was on technological developments and the accomplishments of ice-core drilling projects.

Twenty-nine registrants attended the Symposium, representing Australia, Britain, Canada, France, Iceland, Japan, Switzerland, U.S.S.R., U.S.A., and West Germany. A list of the registrants is included in the volume. The fifteen papers on the program plus one added later, are included here.

The Symposium was under the direction of Dr. Robert H. Rutford, Director of the Ross Ice Shelf Project Management Office, who gave the opening remarks. Dr. Miles Tommeraasen, Vice Chancellor for Business and Finance, University of Nebraska-Lincoln, presented the Welcoming Address for Dr. James H. Zumberge, Chancellor of the University, who was unable to attend. Chancellor Zumberge has had long-time interests in polar glaciology and was the first Director of the RISP Management Office.

The informal arrangement of the Symposium added considerably to its success. In addition to the scheduled papers, a discussion paper was given by Dr. A. Higashi on mechanical properties of ice cores from Antarctica. His abstract is included in the volume. Three films were shown during the Symposium. *Polar Glaciology,* a film produced at the U.S. Army Cold Regions Research and Engineering Laboratory, shows CRREL's activities in that field over the past decade or more. Dr. C.C. Langway, Jr. showed a preliminary film of ice-drilling operations in Greenland in 1974, and Dr. F. Gillet showed a film of the French ice-drilling program.

The editing of the manuscripts was aided considerably by the advice of Mr. B. Lyle Hansen. Technical assistance for the Symposium was capably provided by Mr. Carl K. Cripe and Mr. Karl C. Kuivinen of the RISP office. Mary Swearingen, in addition to her normal duties as RISP secretary, handled commendably the extra load of Symposium organization. Dr. Curt W. Brandhorst, Program Coordinator for the Nebraska Center for Continuing Education, was very helpful in coordinating the accommodations for the Symposium at the Nebraska Center. The volume was typed by Jeanne Bishop. National Science Foundation Grant GV-44390 helped to defray part of the Symposium expenses.

John F. Splettstoesser
Ross Ice Shelf Project

INTRODUCTION

Almost all the drilling and coring in both temperate and cold ice for glaciological and hydrological research has occurred within the last 26 years.

The wide variety of equipment and techniques that have been used can be classified on four bases: whether the means of penetration is thermal or mechanical, the disposition of the meltwater or cuttings formed during penetration, coring or non-coring, and whether hole closure (or opening) due to plastic flow of the ice was controlled. The temperature of the ice, temperate or cold, is an obvious constraint on the equipment and technique that can be used.

Thermal Penetration

Thermal drills have been heated by hot water, steam, burning gases and passing an electric current through resistors.

In temperate ice the meltwater can be left in place. Thousands of meters of hole have been thermally bored in temperate ice; two or more holes reached depths of more than 300 m. Several hundred meters have been cored.

Cold firn can be thermally drilled to the depth at which it becomes impermeable to water without attempting to collect the meltwater. Numerous penetrations have been made to depths of about 50 m.

Thermal drilling or coring in cold impermeable firn or ice has been accomplished by collecting and removing the meltwater; adding an antifreeze to it; displacing it with a heavier non-freezing immiscible fluid; and providing sufficient heat to the hole wall to prevent refreezing.

Mechanical Penetration

Mechanical penetration has been accomplished by rotating a cutting bit, impacting with a chopping bit (cable drilling), vibratory driving of the drill pipe and ballistic impact. For the first two means the motive power may be either a human being or an engine.

Rotary drilling and coring in temperate ice has been accomplished, removing the cuttings by augering or circulating water. For cable drilling at least part of the hole is filled with water to flush the cuttings from the chopping bit and hole bottom. The cuttings are removed by a bailer.

Rotary drilling and coring in cold ice has been accomplished, removing the cuttings by augering, circulating a fluid (either air or a non-freezing liquid), and dissolving the cuttings in aqueous ethylene glycol. Ragle *et al.* (1964) used a SIPRE 3-in. coring auger to core drill to a depth of 55 m on the Ward Hunt Ice Shelf in 1960. This is surely a record depth for man-powered core drilling in ice. The 2164-m-deep hole through the Antarctic Ice Sheet at Byrd Station by the CRREL drill team in January 1968 is the deepest hole drilled in ice to date.

Vibratory drilling and coring in cold ice to depths of tens of meters has been accomplished. With this type of penetration and with the ballistic impact the ice flows away from the penetrator and there are no cuttings to remove.

For those who want to delve into the history of drilling and coring in ice several summaries have been written. The most recent one, by Langway (1970, p. 6 and 146-148), is a small portion of a larger work. Langway's paper has an extensive bibliography citing most of the references pertinent to the history of core drilling in ice from 1949 to 1969. Miller (1952-1953) has covered thermal drilling prior to 1953, and also mechanical drilling in ice prior to 1950 (Miller, 1954).

REFERENCES

Langway, Chester C. Jr., 1970, Stratigraphic analysis of a deep ice core from Greenland: Geological Society of America Special Paper 125, 186 p.

Miller, M.M., 1952-1953, The application of electro-thermic boring methods to englacial research with special reference to the Juneau Icefield investigation in 1952-53: Arctic Institute of North America Report No. 4, Project ONR-86.

Miller, M.M., 1954, Juneau Icefield Research Project, Alaska, 1950: American Geographical Society, JIRP Report No. 7.

Ragle, R.H., R.G. Blair and L.E. Persson, 1964, Ice core studies of Ward Hunt Ice Shelf, 1960: *Journal of Glaciology,* v. 5, no. 37, pp. 39-59.

B.L. Hansen

LIST OF REGISTRANTS

IAN G. BIRD
Antarctic Division
Department of Science
568 St. Kilda Road
Melbourne, 3004, Australia

CARL K. CRIPE
Ross Ice Shelf Project
University of Nebraska
Lincoln, Nebraska 68588

DAVID EILERS
Ross Ice Shelf Project
University of Nebraska
Lincoln, Nebraska 68588

DAVID H. ELLIOT
Institute of Polar Studies
Ohio State University
Columbus, Ohio 43210

FRANÇOIS GILLET
Laboratoire de Glaciologie
2 rue Très Cloîtres
38-Grenoble, France

L.D. GOULD
USA CRREL
Box 282
Hanover, New Hampshire 03755

B. LYLE HANSEN
Ross Ice Shelf Project
University of Nebraska
Lincoln, Nebraska 68588

WILLIAM D. HARRISON
Geophysical Institute
University of Alaska
Fairbanks, Alaska 99701

AKIRA HIGASHI
Department of Applied Physics
Faculty of Engineering
Hokkaido University
Sapporo, Japan

YE. S. KOROTKEVICH
Arctic and Antarctic Institute
Leningrad, U.S.S.R.

KARL C. KUIVINEN
Ross Ice Shelf Project
University of Nebraska
Lincoln, Nebraska 68588

CHESTER C. LANGWAY, JR.
USA CRREL
Box 282
Hanover, New Hampshire 03755

CLAUDE J. LORIUS
Laboratoire de Glaciologie
2 rue Très Cloîtres
38-Grenoble, France

MALCOLM MELLOR
USA CRREL
Box 282
Hanover, New Hampshire 03755

W.S.B. PATERSON
Department of Energy, Mines and Resources
Polar Continental Shelf Project
Room 304, City Centre Terminal
880 Wellington Street
Ottawa, Ontario, K1A OE4,
Canada

KARL PHILBERTH
D 8031 Puchheim/München
Peter-Rosegger-Strasse 6
West Germany

RICHARD PUTNEY
Nebraska Representative,
Mobile Drilling Co., Inc.
4410 Stockwell Street
Lincoln, Nebraska 68506

JOHN H. RAND
USA CRREL
Box 282
Hanover, New Hampshire 03755

GORDON de Q. ROBIN
Scott Polar Research Institute
Cambridge CB2 1ER
England

HEINRICH RUFLI
Physics Institute
University of Bern
Sidlerstrasse 5
3012 Bern, Switzerland

ROBERT H. RUTFORD
Ross Ice Shelf Project
University of Nebraska
Lincoln, Nebraska 68588

JOHN F. SPLETTSTOESSER
Ross Ice Shelf Project
University of Nebraska
Lincoln, Nebraska 68588

YOSIO SUZUKI
Institute of Low Temperature Science
Hokkaido University
Sapporo, Japan

PÁLL THEODÓRSSON
Science Institute
University of Iceland
Reykjavik, Iceland

ROBERT H. THOMAS
Ross Ice Shelf Project
University of Nebraska
Lincoln, Nebraska 68588

LONNIE G. THOMPSON
Institute of Polar Studies
Ohio State University
Columbus, Ohio 43210

RON L. WEAVER
Institute of Arctic and Alpine Research
University of Colorado
Boulder, Colorado 80302

PETER N. WEBB
Department of Geology
Northern Illinois University
Dekalb, Illinois 60115

IAN M. WHILLANS
Institute of Polar Studies
Ohio State University
Columbus, Ohio 43210

CONTENTS

THERMAL ICE DRILLING:

AUSTRALIAN DEVELOPMENTS AND EXPERIENCE

I.G. Bird*
Antarctic Division
Department of Science
Melbourne, Australia

ABSTRACT

During the past seven years the Antarctic Division has gained considerable experience in the thermal core drilling of intermediate depth boreholes. Following CRREL developments, a drill plant especially suited to the requirements of the Australian expeditions has been developed. The new drill is described; it features caravan housing, improved operator convenience and protective devices. Drilling experience and borehole logging instruments are also discussed.

Introduction

Ice drilling has become an important facet of glaciology, particularly for the study of historical changes in the polar ice sheets. Factors such as age, temperature of deposition, particle trajectories, impurities and crystalline properties require the application of ice drilling procedures for investigation. Techniques for the rapid core drilling of intermediate depth boreholes (to 500 m) by thermal methods have been developed over the past decade (Hansen and Langway, 1966; Ueda and Garfield, 1969). The impetus for thermal drill development was mainly the high cost, weight and logistics problems associated with rotary drilling plants (Langway, 1967, pp. 102-104).

During four seasons of field operations over the past seven years the Antarctic Division has core drilled eight intermediate depth boreholes using a thermal drill; more than 1600 m of core has been recovered. Experience with the thermal drilling technique was initiated in 1968 following the acquisition of a drill from U.S. Army CRREL. This drill successfully cored a 310-m hole in the Amery Ice Shelf during that year. The very adverse operational conditions experienced on the Amery Ice Shelf severely stressed both the drill plant and its operators. Under such field conditions, exposure, remoteness from base facilities, minimum staff and limited power made the operation extremely difficult. As future Australian operations in similar situations were intended a modified drill was designed. The new drill features caravan housing and other improvements for operator convenience, including automatic control during drilling, also protective devices, and electronic control of winch speed and drill-head temperature.

---

*Present address: CSIRO, Division of Atmospheric Physics, Melbourne, Australia.

The drill is convenient to operate and reliable in performance. However, several notable failures have occurred including loss of a drill head and cable.

Principles of the Thermal Drill

A practical thermal drilling system comprises a drill head, a means of raising and lowering this head and a cable to transmit power to it. The drill described here melts a hole of 1 cm diameter and takes a core of 11.75 cm diameter and 2 m long. The meltwater is drawn from the drill face by a vacuum pump and is stored in a tank on the drill head (Fig. 1).

Performance of a thermal ice drill is determined by the parameters of the thin layer of meltwater between the heated drill section and the ice. Drill efficiency and speed of penetration are dependent upon the mechanical configuration of the heated annulus, input power and drilling pressure, and the pertinent physical properties of the ice to be drilled, namely temperature and density (Shreve, 1962). Drill penetration rate is typically 4 cm per minute for a heater dissipation of 3.9 kW and an ice temperature of -20°C. Pendulum steering is essential for a plumb hole; this requires a controlled drilling pressure.

With increasing depth of the borehole, a greater proportion of the cycle time is lost in winching the head to and fro between the iceface and the surface. In general, this consideration and borehole closure caused by ice creep limits the practical depth for the present thermal drill in a dry hole to about 500 m.

The 385-m hole drilled at the summit of the Law Dome in 1969 took 21 days to complete on the basis of a two-shift 24-hour day, that is a total of 500 working hours, including maintenance periods. Drilling rates to 44 m per day have been reported by a group drilling at Casey in 1974.

The Antarctic Division Thermal Drill

The basic design objective was to produce a drill for use inland from established Antarctic stations that was self contained, easy to transport and convenient to operate; reliability was paramount for, in general, field operations under our conditions must be self supporting. The general mechanical arrangements of the thermal drill detailed in CRREL drawings series TS66-26 (drill head), TS65-13 (winch assembly) and TS66-34 (winch and tower) were retained. The Australian modifications largely have been described already (Bird and Ballantyne, 1971) but are presented here with the details of latest developments.

General Construction

Environmental protection against the Antarctic weather conditions (low temperatures, high wind and snow drift) is important for both the drill rig and operators.

The drill plant is installed within a caravan 3.7 m by 2.15 m and 2.15 m high; sledge mounting makes for maneuverability. The tower is sealed through the caravan roof and braced to its structure; flexible ducting allows the head to be elevated above the caravan roof for removal of the ice cores. A flexible tube connects the caravan floor to the borehole entrance tube and ex-

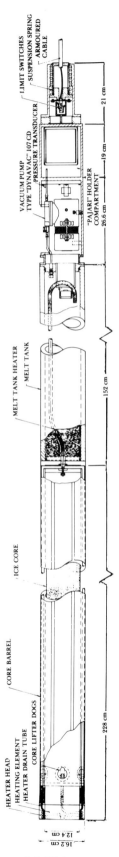

Figure 1. Schematic diagram of the thermal drill (not to scale).

3

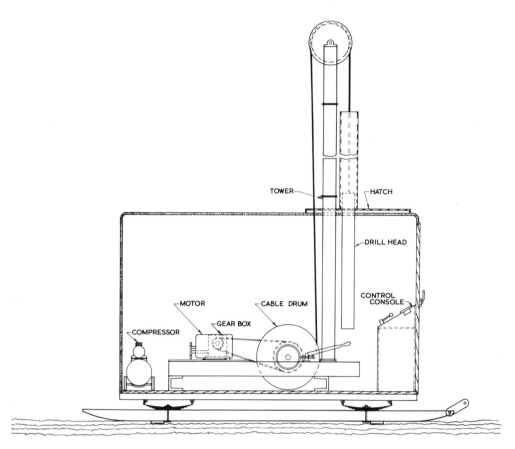

Figure 2. Schematic diagram of drill caravan assembly.

cludes extraneous materials and snow drift. The caravan chassis is shored with timber to support the winch assembly during drilling. Control and monitoring equipment is mounted in a console located adjacent to the drill tower (Fig. 2). A hatch in the caravan roof allows the tower to be lowered, then sectioned and housed in the caravan for transportation.

The caravan is made of 2.5-cm-square steel section, sheeted externally with light-gauge steel and internally with plywood; insulation is included in the wall cavity. Such a structure is light-weight yet is sufficiently strong to withstand difficult traverse conditions. Electrical heating near the control console provides a measure of operator comfort.

The Drill Head

Proper functioning of the drill head ensures a plumb hole drilled at optimum rate. A malfunction can cause delays and accompanying complications due to borehole closure. For the new drill, the drilling process is monitored; thus irregularities are recognized early. Also the design of the heated annulus has been modified to make the consequences of failures potentially less catastrophic.

The Heated Annulus

The efficiency of the drill is determined primarily by the rate at which the heat generated within the drill-head heater is conducted into the annulus and the ice face. It was found that cartridge elements previously used for heating the annulus had a number of undesirable features in practice, for example:

—the fine machining tolerance necessary for assembly of the cartridge units into the annulus;

—susceptibility to failure when immersed in water;

—multiplicity of vulnerable electrical connections;

—high cost;

—difficult to service.

For this reason a single element cast integrally with the aluminum annulus is now used (Fig. 3). It consists of three turns of standard 230-V sheathed heating element; dissipation is 3.9 kW. This arrangement has proved simple to manufacture, also efficient and rugged in operation. The transition area between the heated annulus and the core barrel is filled with a high temperature fiberglass-epoxy resin mix to maintain a uniform exterior profile.

The elaborate system described in an earlier report (Bird and Ballantyne, 1971) which aimed to protect the cartridge heaters and associated wiring from operational damage, proved tedious to manufacture and inefficient in operation. Inefficiencies arose from the thermal resistance which developed between the heated annulus and its surrounding shroud due to oxidation and distortion.

Although it was known that a drill-head heater consisting of a coil of bare resistance wire in direct contact with the ice face had proved highly successful (F. Gillet, personal communication, 1972), the 4-kVA voltage step-down transformer required could not be accommodated readily in the present drill.

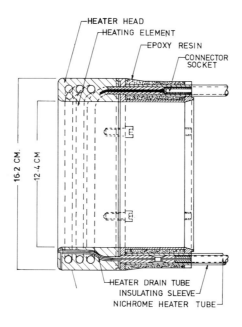

Figure 3. Heated annulus and core-barrel transition section.

Drill-Head Monitoring Systems

Operational experience had demonstrated the importance of monitoring the drilling process, particularly factors associated with the removal and collection of the meltwater. Failure to remove meltwater from the drill face or to detect overflow of the meltwater tank can readily lead to "freeze-in" of the head, an ever-present danger associated with thermal drilling.

(i) *Vacuum monitor:* A melt-tank vacuum gauge has proved an invaluable monitor of drill performance; loss of vacuum, erratic operation of the vacuum pump and melt-tube blockages are readily detected. A solid-state vacuum transducer, "National" model LX1600A, provides a DC output voltage proportional to vacuum level; this information is monitored by a meter at the surface (Fig. 4).

(ii) *Melt-tank water level indicator:* Should the melt tank overflow, the excess water is ejected via the vacuum-pump outlet line; this water rapidly refreezes around and about the drill head. Freeze-in of the head is then inevitable. In the modified drill a water-level transducer is used consisting of a reed relay activated by a float-mounted magnet. The relay grounds the vacuum monitor line when the tank is filled and activates a "Sonalert" audio alarm (Fig. 4). This alarm facility allows maximum-length cores to be taken safely and guards against failure of the operator to empty the melt tank between drilling runs.

(iii) *Heated annulus temperature control:* Under certain operational conditions, such as hot shaving of the borehole, the heated annulus may lose, or maintain only partial, contact with the ice face; serious overheating and damage to the annulus can result. To guard against

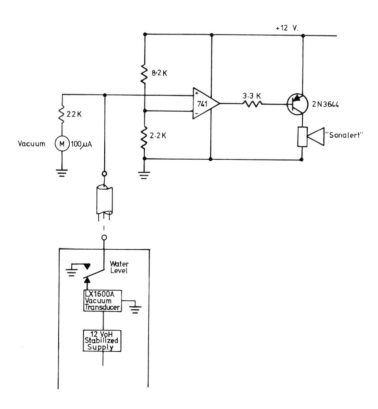

Figure 4. Melt-tank vacuum transducer and water-level alarm circuit.

overheating, thyristor control of the drill-head heater power and monitoring of the annulus temperature are provided.

(iv) *Borehole and core orientation:* It is important to know that a vertical hole is being drilled, also the orientation of the ice core.

A "Pajari" borehole surveying instrument, for the determination of azimuth and inclination, is mounted within the drill head. The instrument monitors the verticality of the borehole, and the orientation of ice cores with respect to the local magnetic meridian; it contains a timer for locking the instrument at a predetermined time during the coring process. Azimuth and inclination are measured separately by alternative orientation of the "Pajari" instrument.

A stylus mounted into the heated annulus grooves a reference line along each core; the line is easily related to the azimuth measured by the "Pajari" and forms a permanent record of core orientation in the drill head. The stylus retracts and does not hinder removal of the core.

Winch and Control Equipment

The winch is powered by a variable-speed, reversible electric motor; speed control over a maximum range of 20-1 is achieved by thyristor control of the motor armature voltage. Electronic control allows a wide range of speed at rated torque, forward and reverse operation with dynamic braking, overload protection, and high reliability.

The average winch load is about 300 kg (drill head 90 kg and say 350 m of cable of weight 200 kg). On the basis of this load and the available electrical power, a 1.5-kW motor was selected. This thyristor-controlled motor at full load draws 2.7 kVA from the single-phase 230-V 50-Hz supply giving a maximum winching speed to the surface of about 20 m/min. For return of the head to the ice face, the free-wheeling speed is limited by braking, for safety.

An air-activated control is provided for the winch brake and clutch. This adds to the complexity of the original CRREL mechanical system but makes a worthwhile advance in operational convenience by allowing centralizing of the drill controls on a console. The variable-speed motor is coupled to the winch drum with chain-sprocket drives via the air clutch and a 40-1 worm-gear reducer. An air-operated dog-clutch disconnects the winch drum from the gear reducer for free-wheeling.

Control Console

All controls and monitors are housed within a single control console. These include controls to the air clutch, air brake, winch motor and drill-head heater, also power metering, air control and drill monitors. The operator thus has full command of all the essential functions at one convenient location (Fig. 5).

Air-activated Control of the Winch

The air clutch connects the motor to the winch drum gear reducer; a variable hand-controlled air valve allows smooth application of winch power via the clutch. Generous braking

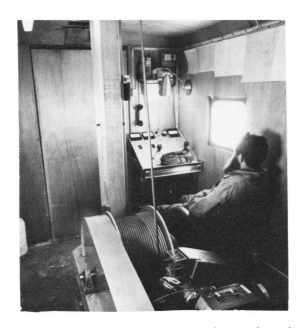

Figure 5. Ice drill in operation, showing the winch, control console and operator.

capability is provided by a 25-cm disc brake; an emergency hand brake is within easy reach of the operator should the air supply or disc brake mechanism fail.

Automatic Drilling Techniques

To drill a plumb hole at optimum rate, about 20 per cent of the drill-head weight should bear on the ice face.

The original CRREL design achieves this function by a calibrated suspension mechanism (Fig. 6) and a hand brake. The drill head weighs on average 90 kg and the suspension spring sensitivity is 8 kg/cm; two limit switches indicate the percentage of drill weight bearing on the ice face. The range 8-23 per cent is indicated by a green lamp, activated by the upper limit switch; range 16-30 per cent operates a red lamp via the lower limit switch. The winch drum was braked by hand to maintain the required drilling pressure, that is just within the red light end of the green light range (Ueda, 1966).

Manual drill feed proved tedious in practice, and with electronic control of the winch motor it became feasible to introduce an automatic servo-controlled feed system by switching the winch motor armature current with a relay activated by the lower limit switch. The drill head is gently lowered in approximately 5-mm increments, thus automatically maintaining correct drilling pressure for pendulum steering.

The great advantage of such a system is that the operator is free to perform routine core analysis and storage work during the 50-min coring period. Only cursory observation of the drilling is required during this period as the safety features outlined earlier adequately monitor the operation.

8

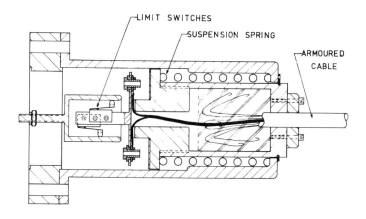

Figure 6. Drill suspension mechanism (also showing cable termination).

An alternative approach to automated drilling is to control the drilling pressure through operation of the air brake by the lower limit switch (Bird and Ballantyne, 1971). However, this technique can cause rapid acceleration and deceleration of the head, and provides a generally less satisfactory descent characteristic than is achieved using motor switching (K. Gooley and D. Russell, Antarctic Division communication from Casey Station, Antarctica, 1974).

Thermal Drilling Experience

Except for the 310-m borehole drilled in the Amery Ice Shelf during 1968, all the Australian drilling activity has been undertaken in the Casey region, on Law Dome (Fig. 7).

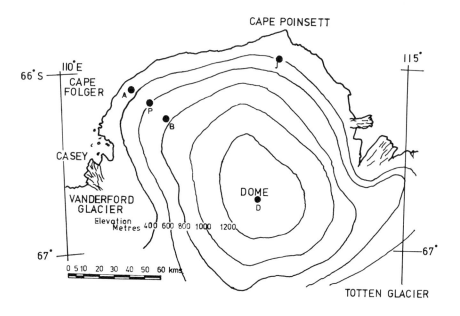

Figure 7. Map of the Casey region, showing the location of main drilling sites.

A summary of drilling operations during the period 1968-74 is given below (Table 1); discussion of these operational activities follows the summary table.

Table 1

Summary of Drilling Operations

Location	Year	Borehole Depth, m	Drilling Period, weeks at site
Amery Ice Shelf:	1968	310	20
—Station G1			
Casey:	1969		
—Cape Folger		324	10
—Law Dome Summit		385	3
Casey:	1972		
—Cape Poinsett		112	4
—Strain grid B		73	1½
—Strain grid P		113	2
—Station S1		53	1
Casey:	1974		
—Cape Folger		348	2
—Cape Folger		in progress	

Amery Ice Shelf

The Amery Ice Shelf is of major interest as a monitor of outflow from the Lambert Glacier system. A segment of the 1968 Amery Ice Shelf project was to core two boreholes with a CRREL thermal drill at points G1 and G3, respectively 67 km and 240 km inland from the ice front.

It took five months overall to drill the 310-m hole at G1, delays being caused by inclement weather and drill malfunctions; loss of time and consumption of spares prevented drilling at G3. Major difficulties were caused by low temperatures, wind and snow drift; the lack of housing round the drilling rig was a significant factor in preventing operation in winds exceeding 12 m/sec. Meltwater pump malfunction caused the drill head to be frozen in at the 60-m level; it was finally recovered after several days of 450-kg cable tension and jerking, the application of several liters of alcohol and intermittent operation of the heaters.

10

The Amery experience clearly demonstrated the need for adequate housing of the drill rig, the desirability of automated drilling, the difficulty in removing unserviceable cartridge heaters from the head, and the need for telemetry of melt–tank vacuum and water level. Rubber vee-belts gave trouble at low temperatures, water and alcohol collected in the unsealed heater relay well causing relay malfunction, and cable twisting produced 200° of rotation over 60 m of drilling.

These problems, while causing consternation at the time, do not detract from the soundness of the CRREL thermal-drill concept. The fact that a borehole was successfully cored under such adverse conditions is a tribute to the drill rig and the tenacity of the four-man Amery team; closure finally prevented continuation of the drilling operation. In general the core quality was good, although there were problems due to shattering when cores were exposed to the thermal shock of -40°C and 10 m/sec winds. About 40 per cent of the cores were returned to Australia for analysis.

Law Dome (Casey Area)

Since 1959, the Antarctic Division has been actively involved in glaciological study of Law Dome; the program includes thermal drilling at various locations (Fig. 7).

1969 Program

A drill of modified CRREL design was constructed during 1968 and successfully cored two holes in 1969. Core recovery from both boreholes was generally satisfactory, although a good deal of fracturing occurred. About 20 per cent of the core was returned to Australia for analysis, the remainder being stored near the drill sites.

Cape Folger was the first drilling site; work commenced on 19 May 1969 and concluded on 13 August. The program was delayed at the 150-m level from 12 June to 4 July due to illness of one of the four-man drilling party. Drilling proceeded normally to 324 m, where borehole closure became a problem and hot shaving was necessary to maintain drill clearance. This process, however, may have enhanced the creep rate, and eventually the drill made a new route at 232 m. Borehole closure could possibly also have been caused by horizontal shear of the hole. In general, the new drill performed well during this operation, most of the problems experienced at Amery Ice Shelf having been overcome. The automatic drill feed, the caravan housing and centralized controls were notably successful. Figure 8 shows an ice core being removed from this drill.

The Dome summit hole was commenced on 28 October and very good progress was made to 385 m in the ensuing three weeks. It took one hour to drill each core, using the automatic feed system, at an ice temperature of about -21.5°C. By 22 November both the melt-tank vacuum and water-level transducers needed, but were not given, maintenance. The drill became frozen in on 23 November.

Chronological events leading to the loss of the drill head and cable are listed below:

November 23 $-$0820 hours: At a depth of 385 m there were indications of probable trouble with the drill and the head was winched to the surface at 0900 hr. The pump flap valve was found open and thus meltwater was not collected for all or part of the period.

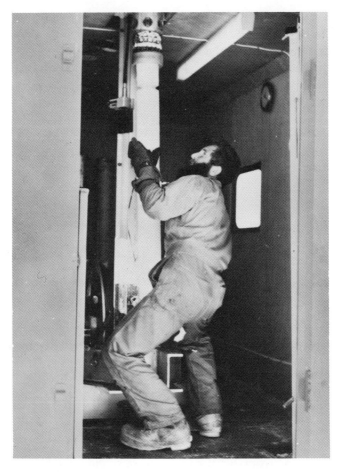

Figure 8. An ice core being removed from the thermal drill.

       −0926 hours: The head was returned to the ice face for one hour of coring; the operator neglected to empty the previously collected meltwater.

       −1010 hours: Trouble was suspected; the head could not be winched and was apparently firmly frozen into position. Cable tension to 2300 kg was applied immediately and 65 liters of alcohol was poured down the hole; the cable was vigorously jerked. The heaters were operated intermittently during the next 36 hours.

Review of these circumstances suggests that the melt tank overflowed soon after 0926 hr and meltwater was ejected over the top of the drill head.

November 25: A further 45 liters of alcohol was applied at 1030 hr and 1130 hr but there was no sign of loosening; the heaters short-circuited at 1050 hr. The drilling party then returned to Casey for additional supplies of alcohol.

December 4: 110 liters of alcohol was applied and the cable tensioned to 4000 kg periodically during the next 6 days.

December 10: There was no indication of movement in the drill head, and reluctantly the decision was made to cut the cable.

December 14: Explosive charges, comprising quarter-sticks of gelignite, were attached to the cable and dispatched to the base of the borehole. Six such charges appeared not to ignite; the seventh, however, jammed 15 m down the hole and severed the cable at this point. Thus, both the drill head and 370 m of cable were lost and the presence of the cable in the borehole prevented further measurement.

This series of events demonstrates the importance of drill-head telemetry, the likely consequences of failure to utilize it properly, and the need for effective cable-cutting equipment.

1972 Program

The principal aim of the 1972 program was to continue the core drilling studies by extending the coverage more widely over the northern survey triangle of Law Dome (Fig. 7, D, J, A). The program included drilling at strain grid J to bedrock, and H, G and B to at least 200 m.

(i) *Cape Poinsett:* Drilling commenced at Cape Poinsett (strain grid J) on 8 March. Beyond 30 m depth drilling speed slowed to 1 m per hour and core diameter decreased; at 50 m a partial failure was thought to have occurred in the winch speed control system reducing the torque output by 40 per cent. The cause of this failure was not established and inadequate winching power continued to retard the drilling operation although the 1.5-kW winch motor was obviously equal to the task. Continuous bad weather and minor malfunctions of the drill also caused problems; following one month of intermittent operation a depth of 112 m was reached. Borehole closure and curvature finally forced the abandonment of the site.

(ii) *Strain grid B:* Following overhaul of the drill plant at Casey during the midwinter period, drilling was recommenced at strain grid B on 23 August. The drill performed well; however, at 73 m a block of timber fell into the borehole and could not be dislodged. The borehole was abandoned.

(iii) *Strain grid P:* A borehole to 113 m was successfully cored at strain grid P commencing on 18 September but lack of winch power forced cessation of drilling at that depth.

(iv) *Station S1:* A 50-m borehole was drilled at S1 for future use in trials of carbon dioxide gas extraction equipment.

During 1972 four boreholes were cored and a total of 350 m of core was recovered; winch and heater head problems had precluded a more successful operation. An analysis of these problems during equipment maintenance in Australia revealed the following:

— the worm-gear reducer which couples the motor to the winch had no lubrication and transmission efficiency naturally was extremely low. Following replacement of a badly worn worm and wheel and the addition of correct lubrication, loads of the order 500 kg could be raised (maximum weight of the drill head and 500 m of cable is 450 kg).

— no fault could be found with the electronic speed control equipment.

— the thermal resistance between the annulus and shroud had risen significantly due to wear and oxidation; too little heat was being transferred to the drill face.

13

Law Dome 1974

The 1974 program includes drilling to bedrock (390 m estimate) at a site 3 km upstream of the existing 324-m borehole at Cape Folger, also to drill at an additional site several kilometers further inland.

Drilling commenced on 14 March and reached a depth of 206 m within 10 days. Several minor problems occurred:

— the solid-state vacuum transducer failed because of voltage transients coupled between adjacent conductors within the drill cable; suppression cured this.

— the winch drum shaft failed at a flange weld.

— the insulation of the electrical connector at the heater head failed due to water immersion.

Operations ceased on 24 March pending repair of the winch drum shaft, drilling recommenced on 3 April, and continued normally to 348 m during the following 3 days. Borehole closure was becoming an increasing problem and shaving was commenced. During a traverse of the borehole the head jammed at 300 m and the heated annulus broke away when moderate tension was applied to the cable. Attempts to dislodge the annulus proved unsuccessful.

The heated annulus design shows it attached to the core-barrel transition section by six high-tensile 6.5-mm-diameter bolts tapped 20 mm into the annulus. However, for an unestablished reason short bolts were fitted during the assembly, possibly as a temporary measure, and these did not penetrate the tapped holes in the annulus. Therefore, only frictional forces held the annulus to the drill head and it was readily dislodged under moderate tension.

A further attempt at drilling to bedrock at Cape Folger is currently in progress.

Borehole Logging Instruments

Instruments for the study of borehole deformation and temperature gradient are important adjuncts to the coring operation. The Division's borehole instruments and associated winches, control equipment and instrument rack are housed within a caravan of similar design to that provided for the ice drill. Two multistrand polythene cables are used for telemetry of the borehole data. The cable drums are powered from a variable-speed electric motor via an automotive differential; this arrangement allows drive to either drum by simply braking the other. Slip rings have not been provided on the winch drums and at selected depths the indicating instruments are plugged to the cables. However, there may be requirements for continuous profiling, for example of borehole diameter, and the feasibility of radio-frequency telemetry from the borehole instruments is under assessment. The major advantages of an RF system are that only a single strain cable would be needed and slip rings could be avoided.

Inclinometer

For the determination of borehole movement, precise measurement of inclination changes up to several degrees is needed. Near the bedrock in coastal regions, the ice temperature can approach pressure melting and inclination changes of up to 20° per year may occur. Measurement accuracy is required to be at least 0.1° and preferably 0.01°, although such precision is difficult to achieve repeatedly in a practical borehole because of sidewall ripple and variable diameter.

The inclinometer described here is a 1-m-long brass tube, 6.25 cm diameter fitted with motorized retractable arms for positioning the tube within the borehole, and a "Schaevitz" angle deviation sensor, model LSRP, mounted within the tube. Its two orthogonal gravity referenced sensors have a range of $\pm5^{\circ}$ and a resolution of 1 second (0.00028°). An "Aanderaa" electrically activated magnetic compass, model K700, mounted at the base of the inclinometer, indicates azimuth to an accuracy of approximately 2° (Fig. 9). Output voltage from both the angle deviation sensor and azimuth compass is displayed on a digital voltmeter; the digital resolution limit of the inclination measurement is $\pm0.001^{\circ}$.

To achieve the desired 0.01° *in-situ* measurement accuracy, strict tolerance restrictions must be held on the machining and assembly of the inclinometer positioning system. The arms are driven into position by a DC motor powered from a constant current supply; the motor may therefore be held in a stall mode. A shear pin holds the worm gear to its shaft and allows the arms to be collapsed should the motor fail.

An alternative approach suggested by H. Krebs (personal communication, 1972) is to mount the angle deviation sensor within a long (2-3 m) heavy tube which will tend to lie naturally along the sidewall of the borehole for even small inclinations; the complication of precision retractable arms is then avoided. Such an arrangement is presently under trial at Casey.

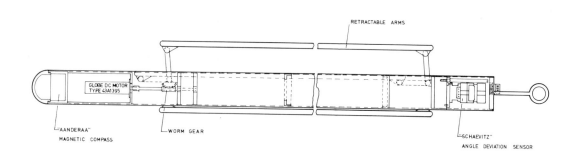

Figure 9. Borehole inclinometer.

Diameter Caliper

For the borehole caliper, three spring-loaded arms provide the positioning force; a measurement resolution of order 1 mm is achieved. The 25-cm arms serve to average the effects of borehole ripple on measurement accuracy. A linear resistance transducer, coupled to a reference voltage, provides a DC voltage output proportional to borehole diameter (Fig. 10). Diameters in the range 7.5-18 cm can be measured.

15

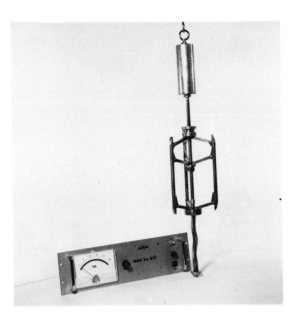

Figure 10. Borehole diameter caliper.

The caliper is fitted with a prototype radio-frequency telemetry system. A voltage-frequency converter allows the caliper DC output voltage to modulate a 100-MHz 50-mW transmitter. The RF signal will be detected at the surface, reconverted to a voltage, and monitored on the digital voltmeter.

Temperature Probe

Borehole temperature measurements, to an accuracy of $0.01^{o}C$, are made using a specially calibrated "Hewlett-Packard" model 2801A quartz thermometer. The temperature sensor must be firmly held against the sidewall of the borehole by an arrangement which does not add significantly to thermal inertia. A protective cap pressed over the sensor end prevents wear due to sliding contact with the borehole.

The temperature probe comprises three lightweight phosphor-bronze spring arms, the sensor being attached to one of these arms by a fingered clamp (Fig. 11). A period of 15 minutes is required for an adequate cooling curve to be plotted and so establish the borehole temperature. The cooling curve is obtained from the quartz thermometer digital output by converting it to an analog voltage and plotting on a chart recorder. End-point temperature may then be readily extrapolated.

Conclusion

The thermal drilling technique has now reached the point where moderately deep boreholes can be drilled almost with certainty and in relative comfort.

The operator may now be well protected from the elements and the automatic feed system

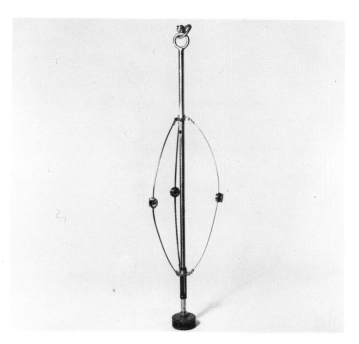

Figure 11. Temperature probe mount.

makes the task far less onerous than in the early stages of development. In addition, the various protective devices which can be incorporated into the drill remove many of the hazards formerly associated with thermal techniques.

Acknowledgments

The major contributions of CRREL in the development of thermal drilling techniques and the ready assistance given to the Antarctic Division through the good offices of Messrs. B.L. Hansen and H. Ueda are acknowledged.

The work of the Amery Ice Shelf Expedition, under very adverse conditions, laid a firm foundation for the continuing work on Law Dome. A number of expedition glaciologists and engineers have contributed to the project: Messrs. M. Corry, N. Collins, A. Nichols (Amery Ice Shelf), and R. Anderson, S. Little, C. Austin, M. Rich, D. Russell, K. Gooley (Law Dome). Mr. J. Ballantyne was responsible for the detailed mechanical design; the drill was constructed at the Division by Messrs. J. Nisbet and W. Kahrau.

The permission of the Director, Antarctic Division to publish this paper is acknowledged.

REFERENCES

Bird, I.G. and J. Ballantyne, 1971, The design and application of a thermal ice drill: Technical Note 3, Dept. of Supply, Antarctic Division, Melbourne, 25 p.

Hansen, B.L. and C.C. Langway Jr., 1966, Deep core drilling in ice and core analysis at Camp Century, Greenland 1961-1966: *Antarctic Journal of the United States,* v. 1, no. 5, pp. 207-208.

Langway, C.C. Jr., 1967, Stratigraphic analysis of a deep ice core from Greenland: U.S. Army CRREL Research Report 77.

Shreve, R.L., 1962, Theory of performance of isothermal solid-nose hotpoints boring in temperate ice: *Journal of Glaciology,* v. 4, no. 32, pp. 151-160.

Ueda, H.T., 1966, CRREL thermal drill instructions: U.S. Army CRREL Report (unnumbered), pp. 5-6.

Ueda, H.T. and D.E. Garfield, 1969, The USA CRREL drill for thermal coring in ice: *Journal of Glaciology,* v. 8, no. 53, pp. 311-314.

A NEW ELECTROTHERMAL DRILL FOR CORING IN ICE

F. Gillet, D. Donnou, and G. Ricou
Laboratoire de Glaciologie CNRS
Grenoble, France

ABSTRACT

The use of a drill tip made of a bare resistance wire fed at low voltage has allowed us to attain drilling speeds of 6 m/hr in temperate ice. The diameters of the hole and the core sample are 140 mm and 102 mm, respectively. In a cold glacier, a vacuum pump eliminates the meltwater. Using an armored electric cable, 500 m long, and a variable speed winch the equipment weighs a total of 1880 kg. The core barrel with the transformer and the unit for removing the meltwater weighs 170 kg and is 8.20 m long. The cores obtained are 2.8 m long.

A new model thermal corer developed at our laboratory was tested for the first time in 1968 on the Saint-Sorlin Glacier, France. It differs from the CRREL model (Ueda and Garfield, 1969) in that it works with a bare resistance wire in direct contact with the ice. This makes it possible to apply a strong power density to the drill tip. A corer making a hole of approximately 14 cm gives a 10.2-cm core, and reaches a speed of 6 m/hr in temperate ice, with 6 kW of heating power. In cold ice, the speed is slightly reduced, and depends on the temperature of ice. The special shape of the heating element has also proved to be very effective for drilling in debris-filled ice, which is often found near bedrock.

Functioning Principles of the Corer

The speed of a thermal drill is determined by the energy density furnished per cm^2 of the cross section of the working face. With a commercial heating element, made of a magnesia-covered resistance, protected by a metal tube, it is impossible to reach a significant power density (Shreve and Kamb, 1964). Moreover, it is difficult to suitably ensure the contact between these heating elements and the metallic mass that surrounds them. This causes still another reduction of the thermal exchanges. On the other hand, by using a bare wire resistance in direct contact with the ice, the power density can be greatly increased (Remenieras and Terrier, 1951). However, the feed rate in the ice is not directly dependent on the dissipated energy per unit of lateral surface of the wire resistance, but rather on the energy density per cm^2 of the drilling cross section. Therefore, it is absolutely necessary that the energy be concentrated in the direction of advance. Technically, the problem was solved by winding a wire resistance into a helix and mounting it on an annular resistance support.

This resistance coil must be able to resist being crushed by the corer, and must not lose its shape, or the sample will be of an unsuitable diameter. (The drill tip is shaped to keep the convecting water currents from going into the core barrel and melting the core.) A wire of sufficient diameter is thus necessary (1.3 - 1.6 mm). Because the resistance of this wire (approximately 2.5 m long) is low (2 - 3 ohms), it is necessary to work at low voltage, with a transformer placed in the corer. Since considerable power is being transported, the on-line losses must be reduced as much as possible. A three-phase current is therefore used. When working in a cold glacier, the meltwater is extracted with a vacuum pump, and stored in a tank that is emptied each time a core is retrieved. The readings of a compass in the borehole are transmitted electrically to the surface, giving the orientation of the ice samples. Finally, a constant tension on the electromechanical cable ensures the verticality of the drilling.

Description of the Installation

The equipment includes the following components and specifications:

(1) *A corer* (Fig. 1), with an outside diameter of 130 mm, that can function as well in a cold glacier as in a temperate glacier when the hole is full of water. In the latter case, the suction assembly for the meltwater is removed. The unit is then 5.9 m long, and weighs 135 kg. In a cold glacier, on the other hand, the total length is 8.2 m and the total weight 170 kg.

(a) *Suspension:* The electromechanical cable is attached with Araldite resin to a piston acting on a spring linked to a linear potentiometer. This gives a reading of the tension that the cable exerts on the corer. At the top of the suspension is a heating resistance of 600 W which can be used in case of wedging on the ascent. A switch at the top enables one to change over to this resistance from the heating resistance of the drill tip.

(b) *Orientation:* An electric motor turns a photoelectric cell around a compass, the disc of which has a hole in it. When the cell arrives in front of the hole, it stops the rotation of the motor. Then, one must simply read the information furnished by the potentiometer connected to the photoelectric cell to find out the orientation of the corer in the horizontal plane. This procedure is, of course, unusable near the magnetic pole.

(c) *Transformer:* Three-phase 338-V primary to 45-V secondary, 6.8-kVA rating, losses 56 W at no load to 460 W at rated load. The transformer is oil-cooled. A balancing piston located at the top allows for free expansion of the oil.

(d) *Recovery of the meltwater:* A vacuum pump lowers the pressure in a tank connected to the drill tip by 3 stainless steel tubes (4 mm I.D., 4.8 mm O.D.). The meltwater is sucked in by these tubes and stored in the tank. To prevent the water from freezing on the way up, the tubes are heated from the inside by a wire with a teflon insulation 1.25 mm in diameter. The normal heating power is 30 W/m. It can be regulated from the surface with a small variable transformer. The heating resistances in the tube and in the reservoir are mounted in series, so the transformer regulates the heating of the whole system. The tank is simply a stainless steel tube, 125 mm I.D. - 129 O.D., and 1.75 m tall. This height corresponds to the amount of water obtained when coring 2.8 m with core 102 - 104 mm diameter and hole 140 mm diameter.

The vacuum pump is a "Marion"-type membrane pump, modified to sit inside the corer. This model has proven to be more reliable and much easier to repair than the Gast model used by CRREL. A monophase 220-V electric motor with a usable power of 120 W

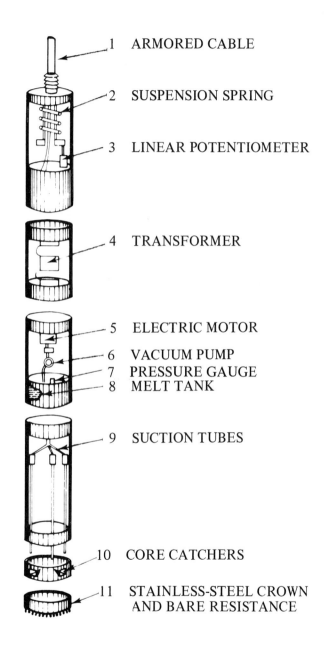

1 ARMORED CABLE

2 SUSPENSION SPRING

3 LINEAR POTENTIOMETER

4 TRANSFORMER

5 ELECTRIC MOTOR

6 VACUUM PUMP

7 PRESSURE GAUGE

8 MELT TANK

9 SUCTION TUBES

10 CORE CATCHERS

11 STAINLESS-STEEL CROWN
AND BARE RESISTANCE

Figure 1. Diagrammatic sketch of corer.

operates the pump. The vacuum obtained can reach 700 mm Hg, which is enough to carry the water up the height of the core barrel. The proper functioning of the suction system can be checked with a vacuum gauge. A level indicator tells when the tank is full.

(e) *Core barrel:* Length 2.8 m, interior diameter 110 mm. It is made of an interior tube of polyethylene (110 mm I.D., 114 mm O.D.) and exterior tube of stainless steel (125 mm I.D., 129 mm O.D.). Between them are the suction tubes for the meltwater, and the electric wires for the resistance at the tip. The stainless steel crown (110 mm I.D., 130 mm O.D.) is hooked on the lower part of the corer. It is tooled at the bottom to allow a porcelain ring housing the bare resistance to be attached. This is attached with six brass screws, three of which are used at the same time for electrical input. Three core catchers are also used for removing the core.

(2) *An electromechanical cable:* 500 m long, 24 mm diameter, weighing 400 kg. The cable contains three conductors with 6 mm^2 cross section used for feeding the head, and nine conductors with 0.75 mm^2 cross section used for feeding the auxiliaries, and for measurements. The insulation is of polyethylene. The stress is supported by a steel braid with a 2200 kg capacity. This braid sits on a strip steel vault which prevents the pull on the braid from transferring the stress to the electrical conductors.

(3) *Winch and mast:* The winch has a chassis of a light alloy on which a movable drum with slip-ring contacts is fixed. The "Lebus" method is used for the winding of the cable, ensuring great functioning security. Until recently a 3 hp motor, connected with a "Jaeger" coupling to a reducing gear, was used for bringing the equipment back up. The purpose of the coupling was to limit the torque in case of jamming, and above all, to allow for an almost automatic descent during the drilling (Bird and Ballantyne, 1971). The free-fall descent was controlled by a foot-operated disc brake. A compressed-air disc brake ensured safety.

This device, after various incidents, has been shown to be unsatisfactory and has just been replaced by a variable-speed motor. Tremendous flexibility is thus obtained at slow speeds (recovery of the core, assembly and disassembly of the corer), as well as at high speeds for the raising and lowering of the unit. The coupling, mounted as a brake, is used only for drilling. The safety brake was also kept. The mast is a duralumin tube 5 mm thick and 200 mm in diameter. It is 8.80 m tall, and comes apart in three pieces. At the top, the pulley is equipped with a pulse generator connected to a depth counter.

(4) *Generator:* The generator provides a three-phase 380-V current. The power of the alternator is 10 kVA, that of the motor 23 hp (the extra power to compensate for losses due to the elevation). Total weight: 220 kg (with starter and batteries).

(5) *Accessories:*

(a) a switch housing with all the controls and dials, including control of the winch. In particular, it contains a variable transformer for controlling the power applied to the drill tip.

(b) for coring done in the Antarctic, the installation is placed in a 5.5 x 3.3 m portable cabin. The walls are made of two sheets of marine plywood (8 mm and 5 mm) enclosing a 15-mm-thick insulation. The whole is held together with aluminum corners.

(6) *Weight of the whole drilling unit* (without crating):

generator	220 kg
corer	170
cable	400
winch, chassis and mast	330
drive platform of the winch drum	180
housing for speed control of winch and brake resistance	290
housing for controls and measurements	190
accessories (compressor, tools)	<u>100</u>
Total	1880 kg

For a drilling operation in the Antarctic, of course, the drilling shelter (600 kg), replacement parts, space for the storage and study of samples, must all be added.

Results Obtained

(1) The first test drilling was carried out in temperate ice in 1968 on the Saint-Sorlin Glacier. Bedrock was reached at a depth of 67 m. The principle of coring with a bare-wire resistance was tested.

(2) In 1969 another coring was performed on the same glacier (Gillet, 1969), down to bedrock at 72 m. The material was then adapted for use in the Antarctic, and for descending to depths of 500 m.

(3) In July 1971, the Vallée Blanche, which, at 3500 m is one of the accumulation basins of the Mer de Glace, was drilled to bedrock at 187 m. The progress of the drilling is shown in Table 1. The drilling speed was fairly high down to 152 m. Various problems, in particular the breaking of the suspension system's potentiometer, which gives the force of the heating resistance against the ice at the bottom of the hole, greatly reduced the speed beyond 152 m. It is worth noting that the power on the heating head was limited to 3300 W (it is possible to go as high as 6000 W). This was done voluntarily to reduce the drilling speed. In fact, an analysis of the free-water content of the ice was performed on each sample as soon as it was extracted. The drilling speed, therefore, had to be adapted to the speed of the analysis. For these three drilling cases, sandy debris, easily observable in the samples, was found in the ice several meters from bedrock. The shape of the resistance proved to be extremely good for drilling through this debris. Sometimes, a small stone crushed several spirals of the helix. In general, these are fairly easy to straighten out. In more serious cases, the resistance must be changed, but this takes only a few minutes. The quality of the samples was excellent.

(4) In January 1972, this same equipment was tested at Terre Adélie [Adélie Coast], Antarctica, several kilometers from the coast. The purpose of this test was to see that the corer functioned properly in a glacier with a negative temperature (here, -15°C). The drilling reached 44 m. Various problems then arose with the vacuum pump motor and with the generator. This campaign permitted us, however, to test the meltwater recovery system, and to determine exactly how big the heating resistance must be to obtain holes of a suitable diameter. It also gave us the opportunity to work under the conditions that prevail on the Antarctic continent. When the granular ice became more compact, and the water remained in the hole, a drilling speed of 4.5 m/hr was obtained. At this time, only 65 per cent of the available power was being used.

Table 1

Progress Chart of the Drilling Operation, Vallée Blanche

Date (1971)	Depth reached, m	Average speed,* m/hr	Average power applied to the head, W	Observations
July 7	32	2.25	1300	Coring in snow at low power
July 8	52.5	2.35	2800	Coring in ice starting at 35 m Breakdown of the generator's regulator
July 9	70	2.3	2800	Repair of regulator Test 4200 W of power; a speed of 5 m/hr is reached
July 10	110	3.6	3300	
July 12	135	3.3	3300	Malfunctioning of the orientation system due to a leak
July 13	152	2.8 1.4	2800	Malfunctioning of suspension's potentiometer. It is then difficult to drill properly causing difficulties in the recovery of the core
July 14	168.5	0.8	2000	Difficulties in the recovery of the core Crushing of the resistance
July 15	180.5	1.6	2000	Fixing short-circuit on the plug of the corer's feeder cable
July 16	187	1.6	2000	Stone trapped in the ice brought up

*The figures are calculated including the time spent in maintenance and minor repairs.

The quality of the samples was very good.

(5) In January 1974, we drilled to a depth of 304 m at Terre Adélie [Adélie Coast], and reached bedrock. The operation, on the whole, went very well. The progress of the drilling is summarized in Fig. 2. A very regular advancement can be seen. It was interrupted only once, at 86 m, by a series of measurements, and a test of thermal reaming of the hole. This reaming was not indispensable, but we wanted to test the possibility of doing it, if necessary. The experiment showed that it is a very delicate operation. In fact, the meltwater cannot be sucked in by the vacuum pump. It runs, therefore, in the hole, and refreezes a little further down. The water that accumulates in this way as ice causes the corer to deviate from its initial path. Therefore, beyond 60 m, because of this deviation, a totally new hole was drilled.

Considering these tests, it seems that if the deformation of the hole makes reaming necessary, it should be done with a mechanical, and not a thermal, procedure.

Despite the relatively high temperatures observed below 200 m (Fig. 3) there were no problems with jamming. As a matter of fact, the corer makes a hole with a diameter of at least 135 mm.

On the other hand, as the drilling below 200 m lasted only three days, the speed of the closing of the hole was insufficient to provoke jamming.

During this drilling, we also tested several different drill tips, with resistances of different diameters and fed with varying powers. Figure 4 represents the drilling speeds obtained, relative to the various powers applied to the tip. A rather large range of results can be seen, which it seems, can only be explained by variations in the quality of the ice. Because of certain deficiencies in the regulation of the generator, it was impossible to feed the drill tip with a power greater than 4100 W.

On the whole, the quality of the samples was excellent. We noticed, however, a significant number of diagonal breaks between 60 and 74 m and 112 to 120 m.

Between 150 and 170 m and from 261 to 264 m the cores had many horizontal fractures. This was also true but in a less noticeable way between 192 and 205 m. Deeper, despite a very strong stratification, the quality of the cores was excellent, and very few fractures were found.

Perspectives

After the excellent results obtained during the coring carried out at Terre Adélie [Adélie Coast] in 1974, it seems very possible that we will be able to drill to depths significantly greater than 500 m, as long as the ice is cold enough to prevent the hole from closing too quickly. We hope, therefore, to drill to a depth of 1000 m during the summer expedition of 1975-1976, in the Dome C region (approx. 74°S, 125°E), located on the Dumont d'Urville-Vostok route.

[Editor's note: This paper was supplemented by a 10-minute color film showing ice-drilling operations in Vallée Blanche.]

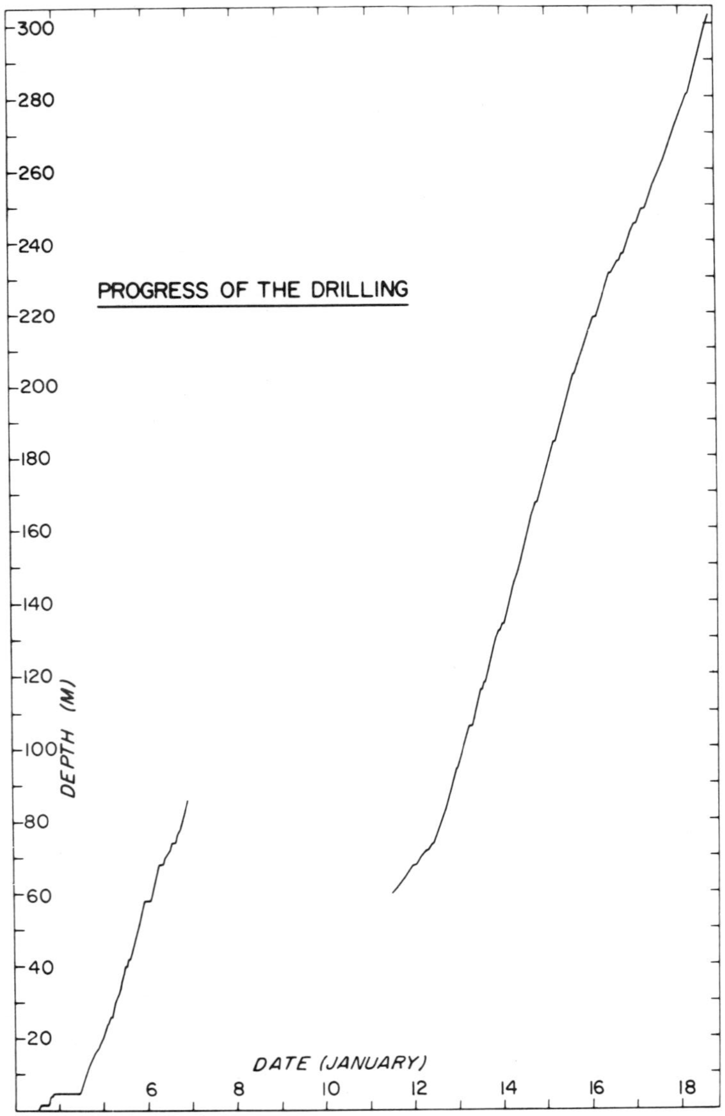

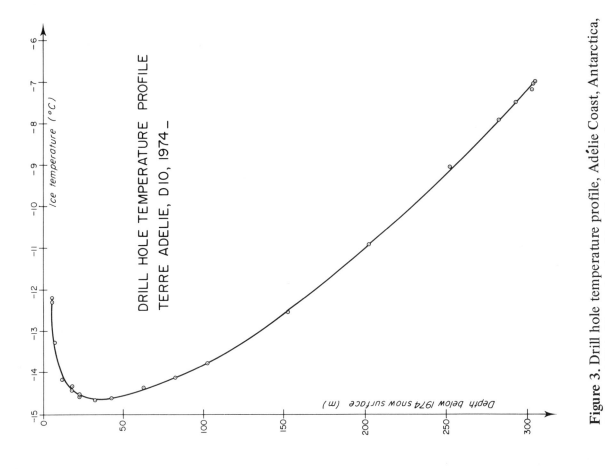

Figure 3. Drill hole temperature profile, Adélie Coast, Antarctica, January 1974. (Data obtained by Claude Rado.)

Figure 2. Progress of drilling, Adélie Coast, Antarctica, January 1974.

26

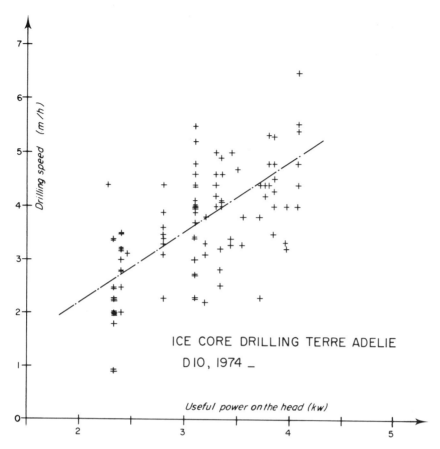

Figure 4. Speeds obtained in ice drilling, Adélie Coast, Antarctica, January 1974.

REFERENCES

Bird, I.G. and J. Ballantyne, 1971, The design and application of a thermal ice drill: Technical Note 3, Dept. of Supply, Antarctic Division, Melbourne, 25 p.

Gillet, F., 1969, Forages et carottages dans le glacier de St-Sorlin: Société Hydrotechnique de France, Section Glaciologie (unpublished).

Remenieras, G. and M. Terrier, 1951, La sonde électro-thermique EDF pour le forage des glaciers: Association Internationale d'Hydrologie Scientifique. Symposium de Bruxelles, août 1951, pp. 254-260.

Shreve, R.L. and W.B. Kamb, 1964, Portable thermal core drill for temperate glaciers: *Journal of Glaciology,* v. 5, no. 37, pp. 113-117.

Ueda, H.T. and D.E. Garfield, 1969, The USA CRREL drill for thermal coring in ice: *Journal of Glaciology,* v. 8, no. 53, pp. 311-314.

DEEP CORE DRILLING IN THE EAST ANTARCTIC ICE SHEET:

A PROSPECTUS

B. Lyle Hansen
Ross Ice Shelf Project
University of Nebraska
Lincoln, Nebraska 68588

ABSTRACT

A major objective of the International Antarctic Glaciological Project (IAGP) is to core drill through the East Antarctic Ice Sheet and into the bedrock beneath it.

The drilling site will be remote from major bases at an elevation of over 3000 m. All equipment and supplies will be flown in using ski-equipped C-130 aircraft.

A unique drill pipe consisting of lengths of fiberglass-reinforced epoxy pipe cemented to lightweight steel tool joints has been developed which weighs only 2.87 kg/m and costs less than $25/m.

This development makes it possible to use a lightweight wireline core-drilling system which minimizes the logistics burden, the time required for drilling, and the cost of the operation.

The wireline core-drilling system consists of a coring bit attached to the core barrel outer-tube assembly which is rotated by the drill pipe, the non-rotating core barrel inner-tube and core-lifter assembly, a wireline hoist with an overshot attached to its cable which is used to retrieve the core-laden inner tube through the inside of the drill pipe, a means of supporting and rotating the drill string, and a means of circulating the drilling fluid which removes the cuttings from the hole and prevents its closure by plastic flow of the ice due to the overburden pressure.

Cold air will be the drilling fluid for the upper 1000 m of the hole. Use of a reverse air-vacuum circulation system eliminates the requirement for an air-cooling system and results in the production of very clean uncontaminated ice cores.

At a depth of about 1000 m arctic-grade diesel fuel (DFA) will be used as the drilling fluid to reduce hole closure and remove the cuttings.

The DFA will be pumped through the drill pipe and carry the cuttings up through the annulus to a separator on the surface. The clarified DFA will be recirculated through the drill string.

It is intended to field test all components of the wireline core drilling system on the Ross

Ice Shelf Project in Antarctica during the 1975-1976 austral summer.

A program is recommended which would make it possible to begin the U.S. deep core drilling in ice portion of the IAGP as early as the 1976-1977 season.

Introduction

One of the objectives of the International Antarctic Glaciological Project (IAGP) (Anonymous, 1971) is to core drill through the East Antarctic Ice Sheet and into the bedrock beneath it.

The technique that was used successfully to core drill through the Greenland Ice Sheet at Camp Century (Ueda and Garfield, 1968) and through the Antarctic Ice Sheet at Byrd Station (Ueda and Garfield, 1969a) will be reviewed briefly to point out why it cannot be used in East Antarctica.

A description of the wireline core-drilling system for the East Antarctic drilling with particular emphasis on the new composite drill pipe and its impact on equipment, financial and logistical requirements is followed by recommendations for the drilling program.

Review of Prior Deep Core Drilling in Ice

It is not possible to drill deep into the ice unless the hole is filled with a fluid to prevent or greatly reduce hole closure due to the flow of the ice caused by the overburden pressure. The hole cannot be filled with fluid unless the hole through the permeable firn which overlies the impermeable ice is lined with a casing frozen into the impermeable ice.

At both Camp Century and Byrd Station the hole for the casing was drilled with a CRREL thermal drill (Ueda and Garfield, 1969b; Ueda and Hansen, 1967).

The fluid used to prevent hole closure was a mixture of trichlorethylene (TCE) and diesel fuel, arctic (DFA) whose density is nearly the same as ice.

An electromechanical coring device (modified Electrodrill) using a hoist and an armored cable to transmit power and to raise and lower the drill was used to core drill from the bottom of the casing to the bottom of the ice sheet and on into the material beneath the ice sheet.

The drill cuttings were removed from the hole using a unique glycol-dilution technique to put them into solution. A volume of concentrated aqueous ethylene-glycol solution, the amount depending upon the downhole ice temperature and the volume of cuttings to be formed, was sent down on each run.

The downhole pump and motor in the Electrodrill caused this fluid to flow through an annular space between the motor and its cylindrical housing on its way to the cutting bit. Heat from the motor transferred to the glycol solution provided the heat needed to dissolve the ice chips.

The diluted solution was removed in the bailer section on each return trip. Any dilute solution remaining downhole stayed downhole because it was denser than, and immiscible with, the mixture of DFA and TCE above it.

It will not be possible to use this technique in East Antarctica where the temperatures in the borehole will be in the minus 50° C range because the fluid is too viscous and the permissible dilution is about zero (Fig. 1).

Wireline Core-Drilling System

A unique wireline core-drilling system is being developed which utilizes components and techniques from both the diamond core drilling and rotary drilling industries and is intermediate in size between the rigs typical of those industries.

It is an outgrowth of the system designed to drill the core and access holes for the Ross Ice Shelf Project. All components except a larger mast and all techniques will be field tested on the RISP.

A wireline core-drilling system consists of a coring bit attached to the core barrel outer-tube assembly which is rotated by the string of drill pipe, the non-rotating core barrel inner-tube and core-lifter assembly, a wireline hoist with an overshot attached to its cable which is used to retrieve the core-laden inner tube through the inside of the drill pipe, a means of supporting and rotating the drill string, and a means of circulating the drilling fluid which removes the cuttings from the borehole, cools the coring bit, and stabilizes the hole.

The objective in core drilling is to obtain a meter of good quality core for every meter of hole drilled and to do it as rapidly as possible.

The rate at which the hole is drilled is a function of four factors: the design of the bit, its rotating speed, the weight on the bit, and the removal of the cuttings by the circulation of the drilling fluid.

Mellor (1974) has examined the design of drag bits and the kinematic relations to rotating speed and the rate at which the hole is drilled. His results have been used to check the design of the bits used on prior deep core drilling and the similar bits designed for the wireline coring system.

These are drag-type coring bits with steel or diamond-set blades designed to cut 60-mm-diameter core in holes either 159 mm or 178 mm diameter. The latter is required to case the upper part of the borehole. The steel-bladed bits are used for coring in firn or ice and the diamond-set bits for coring in the sub-ice material.

Prior experience has shown that rotational speeds of 60 - 225 rpm produce good core. The Bowen S-1E Power Swivel which will be used to support and rotate the drill string on the wireline system has a speed range of 0-150 rpm. Its output torque is adjustable over the range 0-150 m·kg.

The weight on the bit in this wireline system is subject to several constraints. The weight on the bit required to cut ice or the sub-ice material is provided by gravity loading—a portion of the weight of the core barrel and/or drill collars is applied to the bit. The remaining portion keeps

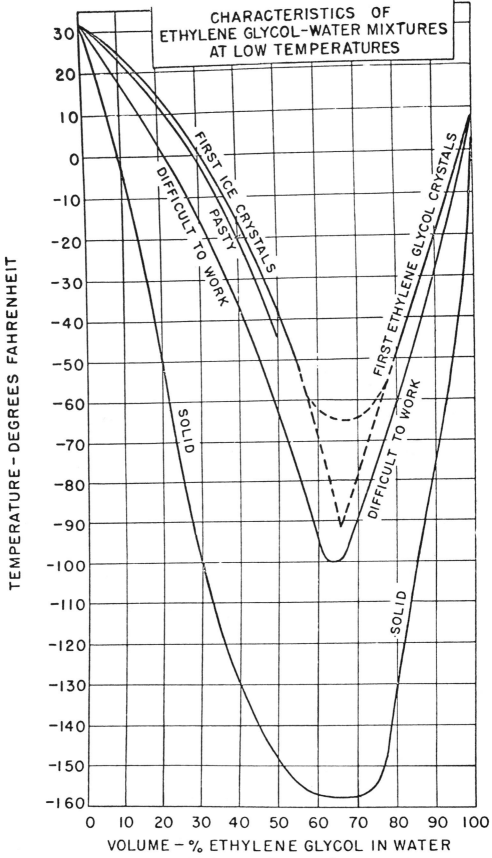

Figure 1. Characteristics of ethylene glycol-water mixtures at low temperatures.

32

the string of drill pipe in tension at all times. The portion of the weight of the core barrel and/or drill collars supported by the drill pipe provides pendulum steering to keep the hole plumb.

The weight on the bit is controlled by adjusting the rate of penetration by means of a flow control valve on a hydraulic cylinder. The entire weight of the drill string is supported by this cylinder which serves a dual purpose: it provides the drilling feed (range 0-0.6 m/min) and it serves as the main hoist with a speed range of 0-30 m/min. Core will be obtained in 6-m lengths.

The cuttings are removed from the hole by circulating cold air or DFA.

Reverse air-vacuum circulation has been used to drill large-diameter holes for several purposes including mine rescue. There are no known instances of its use for core drilling. A vacuum pump connected to the drill pipe through the power swivel causes air to flow down the annulus between the drill pipe and the hole wall, pick up cuttings as it flows past the bit, and carry the cuttings up the drill pipe.

Reverse air-vacuum circulation will be used to drill a 178-mm-diameter hole through the firn and into the impermeable ice so that this portion of the deep hole can be cased. Six-inch (162-mm I.D.) fiberglass-reinforced epoxy pipe will be used as casing. The lower 10 m of the casing and the annulus between it and the hole wall will be filled with water which, when frozen, forms the seal between the casing and the ice.

At the low temperatures in the East Antarctic Ice Sheet it should be possible to use the reverse air-vacuum circulation to core drill to a depth of 1000 m.

At that depth hole closure becomes a problem and it is necessary to fill the hole with DFA.

The DFA will be pumped down through the drill pipe and carry the cuttings up through the annulus to a separator on the surface. The clarified DFA will be recirculated through the drill string.

Circulation of the cold DFA may make it feasible to freeze the interface between the ice and the sub-ice material and prevent the intrusion of water if the bottom of the ice sheet is at the pressure melting point.

Composite Drill Pipe

The weight of the drill pipe required for core drilling to depths of 3000 or more meters in ice is a major factor in the design of the drilling system, the logistics requirements for deployment to and from the drill site, and the cost of the equipment and its utilization.

Core drilling in ice places less stringent requirements on the drill string in the following ways. The ambient temperature is low and many materials are stronger and have higher fatigue resistance at lower temperatures. The cuttings and hole wall are not abrasive. The power required to cut the ice is very low.

Sellmann and Mellor (1972) have examined the power required for drilling in ice. Using their results it is estimated that the power required for drilling a 159-mm-diameter hole at 0.3 m/min is nearly 750 W. The torque at 100 rpm would be 7.2 m·kg.

In 1972, the Longyear Company entered into a feasibility study to determine if it was possible to develop an aluminum-and-steel composite rod within the dimensional confines of their existing PQ Wireline System.

This study indicated that such a rod could be produced and the program progressed through the prototype-manufacture and laboratory-test stages. Laboratory tests have confirmed that a rod manufactured within the size limitations of the PQ System and using aluminum tubes with alloy steel ends would be suitable for deep core drilling in ice.

In November 1972 CRREL undertook the design and development of an even lighter composite drill pipe consisting of CIBA Geigy 512 3 in. (76.2 mm) fiberglass-reinforced epoxy pipe (FRP) cemented to Hydril flush joint wash-pipe (FJ-WP) connections made from N-80 steel. This construction is feasible because the epoxy pipe has almost the same thermal coefficient of expansion as the steel connections.

Over 500 m of this composite pipe has been procured for use on the RISP.

Table 1 is a compilation of data that is useful in evaluating several different configurations of drill pipe. Since each of the drill strings are of different size and vary considerably in construction, material specifications, investment costs, etc., it should be understood that a true comparison of the strings cannot be made. The tables are intended to provide the technical data required to more intelligently plan a program of deep-ice drilling.

The pipes are: (1) Hydril "A-95" tubing connection on modified API upset tubing in Grade "N-80" steel with a nominal outer diameter of 4 in. (101.6 mm) weighing 11 lb/ft (16.4 kg/m), (2) Reynolds 4 in. Aluminum Drill Pipe, (3) E.J. Longyear's composite aluminum drill rod PQ-S, and (4) the CRREL FRP composite.

Comparisons of the weights in kg/m in air—15.83, 14.40, 6.32 and 2.87, respectively—show the drastic reductions achieved by Longyear and CRREL.

Comparisons of the 1973 cost figures in dollars per meter—20.33, 43.62, 82.00 and 24.86, respectively—show the significant saving in the cost of the pipe alone that resulted from the CRREL development. The overall saving that comes from reduction in the size of the drilling equipment required and the transportation costs is much greater.

The total weight of drill pipe and drill collars required for the 30-cm-diameter access hole through the Ross Ice Shelf is 3220 kg. The present drilling mast is designed for a load of 6800 kg.

Core drilling through 3500 m of ice will require a mast that will support 8400 kg.

Only two significant changes need to be made on the RISP wireline core drilling equipment to increase its capability to that required for deep core drilling in East Antarctica: (1) provide a taller mast with a greater weight capacity, and (2) a new wireline hoist designed to work to 3000-4000 m instead of the present 1000 m.

Table 1

Drill Pipe Data

		Hydril A-95	Reynolds Aluminum	Longyear PQ-S	CRREL Composite
Pipe Elements					
Area, A	m^2	1.985×10^{-3}	3.487×10^{-3}	1.898×10^{-3}	1.095×10^{-3}
Moment of Inertia, I	m^4	2.248×10^{-6}	3.992×10^{-6}	2.813×10^{-6}	1.074×10^{-6}
Section Modulus, S	m^3	4.425×10^{-5}	7.486×10^{-5}	4.923×10^{-5}	2.324×10^{-5}
Radius of Gyrations, r	m	3.366×10^{-2}	3.383×10^{-2}	3.851×10^{-2}	3.132×10^{-2}
Polar Moment of Inertia, J	m^4	4.495×10^{-6}	7.985×10^{-6}	5.626×10^{-6}	2.149×10^{-6}
Polar Section Modulus, Z	m^3	8.849×10^{-5}	1.497×10^{-5}	9.845×10^{-5}	4.649×10^{-5}
E I Value	$m^2 kg$	4.741×10^{4}	2.985×10^{4}	2.098×10^{4}	0.221×10^{4}
Dimensions and Weights					
Pipe Body Outside Diameter	m	1.016×10^{-1}	1.067×10^{-1}	1.143×10^{-1}	0.924×10^{-1}
Pipe Body Inside Diameter	m	0.883×10^{-1}	0.833×10^{-1}	1.032×10^{-1}	0.846×10^{-1}
Pipe Body Wall Thickness	m	0.665×10^{-2}	1.168×10^{-2}	0.556×10^{-2}	0.394×10^{-2}
Joint Outer Diameter	m	1.096×10^{-1}	1.460×10^{-1}	1.206×10^{-1}	1.016×10^{-1}
Joint Inner Diameter	m	0.862×10^{-1}	0.826×10^{-1}	1.032×10^{-1}	0.883×10^{-1}
Overall Length Shoulder to Shoulder	m	9.144	9.144	6.096	6.096
Total Volume	m^3	1.847×10^{-2}	3.710×10^{-2}	1.290×10^{-2}	0.713×10^{-2}
Weight with Joints					
In Air	kg/m	1.583×10^{1}	1.440×10^{1}	6.428	2.872
In 1200 kg/m³ mud	kg/m	1.342×10^{1}	9.032	3.898	1.458
In 800 kg/m³ DFA	kg/m	1.422×10^{1}	1.071×10^{1}	4.762	1.934
Joint Properties					
Torsional Yield	m·kg		5.184×10^{3}	1.244×10^{3}	0.484×10^{3}
Make up torque	m·kg	0.484×10^{3}	1.548×10^{3}	0.346×10^{3}	0.152×10^{3}
Tension, min. yield	kg	1.061×10^{5}			6.033×10^{4}
Pipe Properties					
Tension, min. yield load	kg	1.116×10^{5}	1.422×10^{5}	0.774×10^{5}	1.820×10^{4}
Tension, min. yield strength	N/m^2	5.518×10^{8}	4.000×10^{8}	4.000×10^{8}	2.069×10^{8}
Torsional yield strength	m·kg	2.696×10^{3}	3.260×10^{3}	2.143×10^{3}	1.940×10^{2}
Torsional stiffness	kg/rad	3.910×10^{5}	2.418×10^{5}	1.701×10^{5}	0.107×10^{5}
Resistance to collapse	N/m^2	6.070×10^{7}	7.242×10^{7}	3.449×10^{7}	1.690×10^{7}
Internal yield pressure	N/m^2	6.325×10^{7}	7.656×10^{7}	3.221×10^{7}	
Internal burst pressure	N/m^2	8.698×10^{7}	8.449×10^{7}	3.559×10^{7}	3.932×10^{7}
Thermal coefficient of expansion	$°C^{-1}$	1.242×10^{-5}	2.160×10^{-5}	2.160×10^{-5}	1.242×10^{-5}
Stretch per meter per kg load	m	2.388×10^{-8}	3.848×10^{-8}	7.070×10^{-8}	4.811×10^{-7}
Cost, dollars per meter (1973)		20.33	43.62	82.00	24.86

Recommendations

The following program should result in successful deep core drilling in East Antarctica with the least expenditure of funds for equipment, drilling and logistical support.

1. Test the RISP wireline core drilling equipment in Greenland on the GISP-75 program using only the reverse air-vacuum circulation to remove the cuttings. This eliminates the need to case the hole through the firn and would permit core drilling to 400 or 500 m at Jarl-Joset or some other site. It would provide field experience with the equipment and an opportunity to correct any minor deficiencies prior to shipment to Antarctica later in 1975.

2. Proceed with the procurement and construction of wireline core drilling equipment and drill pipe to core drill a hole through the Greenland Ice Sheet in the summer of 1976. This equipment can also be used for intermediate depth holes at other locations in Greenland.

3. Execute the first season's RISP drilling program using the existing equipment during the 1975-1976 season. This program should include the procurement of another 300 m of drill pipe to permit core drilling the sub-sea sediments to a depth of about 30 m which can be accomplished with the present equipment.

4. Procure the larger mast, a 3500-m wireline hoist and 3000 m of drill pipe for use in East Antarctica. These can be used in conjunction with the RISP equipment to carry out the U.S. portion of IAGP deep core drilling in ice program beginning as early as the 1976-1977 season.

Prior experience has demonstrated the desirability of testing drilling equipment and techniques in Greenland prior to their use in Antarctica, a practice which has been followed without exception on all U.S.A. core drilling in ice projects in Antarctica.

REFERENCES

Anonymous, 1971, International Antarctic Glaciological Project: SCAR Bulletin No. 38, pp. 807-811. Reprinted from *Polar Record,* v. 15, no. 98, pp. 829-833.

Mellor, Malcolm, 1974, Kinematics of drag bits for rotary drilling: U.S. Army CRREL Technical Note.

Sellmann, Paul V. and Malcolm Mellor, 1972, Power requirements for drilling in frozen earth materials: U.S. Army CRREL Technical Note.

Ueda, Herbert T. and Donald E. Garfield, 1968, Drilling through the Greenland ice sheet: U.S. Army CRREL Special Report 126.

Ueda, Herbert T. and Donald E. Garfield, 1969a, Core drilling through the Antarctic ice sheet: U.S. Army CRREL Technical Report 231.

Ueda, Herbert T. and Donald E. Garfield, 1969b: The USA CRREL drill for thermal coring in ice: *Journal of Glaciology,* v. 8, no. 53, pp. 311-314.

Ueda, Herbert T. and B. Lyle Hansen, 1967, Installation of deep-core drilling equipment at Byrd Station (1966-1967): *Antarctic Journal of the United States,* v. 2, no. 4, pp. 120-121.

DRILLING TO OBSERVE SUBGLACIAL CONDITIONS

AND SLIDING MOTION

W.D. Harrison* and **Barclay Kamb**
Division of Geological and Planetary Sciences
California Institute of Technology
Pasadena, California 91109†

ABSTRACT

Our ignorance of what happens at the bed of a glacier is the outstanding problem in glacier flow, and serious drilling efforts are necessary for its solution. Some of the problems encountered in a small-scale drilling project to make observations at the bed of a glacier are discussed. These include the problems of thermal and cable-tool drilling and bailing in dirty, actively-deforming temperate ice using back-packable equipment, the problems of borehole photography and water turbidity, and the possible effect of the borehole on the conditions observed.

Introduction

Although considerable progress has been made during the past 20 years in our understanding of the deformation of ice, our ability to predict the flow of a glacier, or to interpret its behavior in terms of climatic changes, is severely restricted by our ignorance of what goes on at the glacier bed. For the same reason, we know very little about the geomorphological processes of glacial erosion and deposition, and some hydrological problems such as water storage in glaciers and the production of sediment in glacially fed streams.

Because ice is a rather plastic material there are conditions under which the thickness of a glacier, and its flow as observed at the surface, are rather insensitive to what is occurring at its bed. Yet the thickness and flow of some other glaciers are probably to a large extent determined by subglacial processes. The list includes the surging glaciers, the large outlet glaciers, possibly some glaciers on volcanoes, and glaciers showing large seasonal variations in velocity.

The importance of the problem is recognized, and considerable theoretical work has been devoted to predicting what should happen at the bed of a temperate glacier; small-scale subglacial topography and water pressure play important roles. But at this stage the problem is essentially

*Present address: Geophysical Institute, University of Alaska, Fairbanks, Alaska 99701.

†Division of Geological and Planetary Sciences, Contribution No. 2540.

an experimental one. Because experiments are difficult, the theory has not been adequately tested, nor has it been shown that the type of interface which it assumes, clean ice lying on bedrock, is necessarily typical. Although some observations have been made in tunnels near terminal, marginal and icefall regions, drilling is the key to observation in the more typical deep regions of a glacier. In fact, one type of borehole experiment is already fairly common. The measured deformation of a borehole, integrated over its length, is subtracted from the measured surface velocity to give the velocity at which the glacier slides over its bed. Failure of the borehole to reach the bed, perhaps not uncommon, results in an overestimate of the sliding velocity.

Some problems encountered in a project to drill through a glacier and make direct observations at the bed with the help of borehole photography are described here.

Drilling

Our experiments were carried out during the summers of 1969 and 1970 in an area of Blue Glacier near the equilibrium line where the ice thickness is about 120 m. Blue Glacier is a temperate glacier on Mt. Olympus, Washington, U.S.A. Most of the drilling was done with 51-mm-diameter electrothermal drills similar to those described by Shreve and Sharp (1970). The technique was similar to that developed by Kamb and Shreve (1966) in that no casing was used. The holes, which tended to refreeze slowly, were maintained at a diameter of 60 mm with a special conically-shaped thermal drill. During the first summer we learned by borehole photography that a thin layer of debris accumulated beneath the thermal drill to a thickness of only 2 or 3 mm was sufficient to slow the thermal drilling rate, normally about 7 m/hr, by perhaps three orders of magnitude.

During the second summer we attempted to penetrate dirty ice with the help of a cable tool (Johnson, 1966). (A cable tool is essentially a heavy chisel which is repeatedly raised some tens of centimeters off the bottom with a cable and dropped.) The string of tools, consisting of swivel, jars, stem, and bit, was about 3 m long and weighed about 30 kg (Fig. 1). It was driven off a cathead, which for convenience was not connected directly to the wire cable, but via a rope clamped to it. The cable was continually adjusted so that there was some tension as the bit struck the bottom. A small cable winch, the cathead, and the sheave were all mounted on a wooden tripod (Shreve and Kamb, 1964, p. 115). We found it advisable to complete the cable-tool drilling fairly promptly. It can be difficult to re-enter a hole with a long string of tools after a period of days or weeks, depending on the rate of deformation of the ice.

The performance of this cable tool was rather poor, as illustrated by a drilling rate of only 1 m/hr in clean ice. We feel that this could have been considerably improved by increasing the 500 W used to power the cathead to a larger fraction of the 2000 W that was available. This is the power used in the electrothermal drilling. Another difficulty was that the bit was probably striking a fairly large area since no casing was driven to guide it, and the hole was usually enlarged by earlier attempts at thermal drilling in the dirty ice. Despite its inefficiency, layers of heavily debris-laden ice were successfully penetrated with this cable tool. It is interesting that a regular well-drilling cable-tool rig was used for experimental drilling in Athabasca Glacier in 1960 (W.S.B. Paterson, personal communication). Both the drilling speed and the weight of the equipment were at least an order of magnitude larger.

Water stood in all the holes in which cable-tool drilling was attempted. This permitted debris to be bailed out with a sand pump (Johnson, 1966). This is a simple device which, like the cable

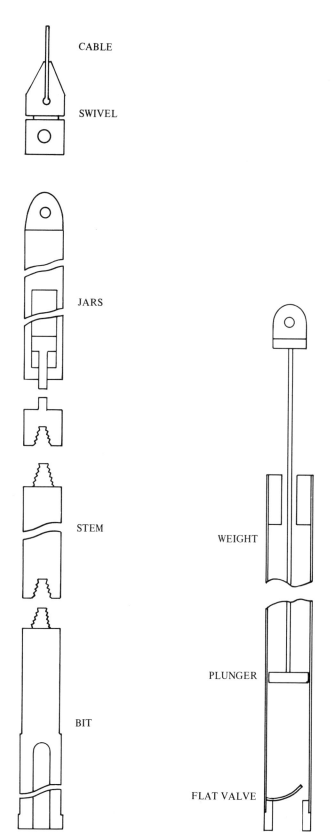

Figure 1. Cable tool (left) and sand pump (right). The bit, stem, and jars are respectively about 0.6, 2.0 and 0.5 m long; the maximum diameter of the bit is about 50 mm. The body of the sand pump is about 1.5 m long and 51 mm in diameter.

tool, is operated with a single cable. The pump is run into the hole with the cable attached to the plunger, which is therefore initially extended. When the pump body reaches the bottom and the cable tension is released, the plunger retracts under the influence of gravity. A strong jerk on the cable sucks debris into the bottom of the pump, where it is contained by a flat valve; we found that a sheet of neoprene worked best for this (Fig. 1).

Our sand pump worked well, except that it seemed to have difficulty in picking up all the debris in enlarged holes; this complicates the determination of undisturbed subglacial conditions. The sand pump seems to be a very useful tool even when no mechanical drilling is attempted. We found that after the thermal drill was stopped by debris, thermal drilling could often be continued after the bottom of the hole was cleaned with the sand pump.

All of the components used in these drilling techniques are portable, simple, and inexpensive. Large depths do not present a great problem because no casing or drilling pipe is used.

Borehole Photography

The boreholes and subglacial conditions were studied with a borehole camera. The instrument and its use have already been described, together with an air-lift system to pump turbid water from the bottom of the hole (Harrison and Kamb, 1973). A sketch of the camera is shown in Fig. 2. Turbidity caused by mechanical drilling and bailing is the most serious problem facing borehole photography. Even after the water pumped from the bottom of our holes became clear, by the time the air lift was run out of the hole and the camera run in (5 or 10 minutes), turbidity often prevented photography of the hole bottom. The problem decreased over a period of days or weeks until pumping immediately before photography was not required. No attempt to use a flocculent was made. We do not yet understand the physical processes involved in this turbidity problem.

Some of the capabilities of borehole photography have also been described (Harrison and Kamb, 1973). Perhaps the most striking is the observation of the motion of the ice with respect to the bed, as illustrated by Fig. 3. The observation of motion like this is probably an indication that the bed has been reached. Photos such as those of Fig. 3 are a stereo pair and therefore contain small-scale topographic information about the bottom, which is an ingredient of glacier sliding theory.

Effects of the Borehole

The presence of a borehole changes both the stress and the temperature environments. Depending upon the magnitudes of impurity and water pressure effects, either melting or freezing can take place (see Harrison, 1972). Other possible disturbances can be imagined. For example, say a borehole intersects the bed on the downstream side of a rock protuberance; this should be a low-pressure region. If the hole is water-filled, and the water does not escape due to surrounding higher pressure regions, there will be an excess pressure at the hole bottom which may cause the ice to separate from the bed. Disturbances such as these are probably not serious for most observations, as long as they are made immediately upon completion of the borehole. This points out the need to be able to handle the turbidity problem, which is most severe then.

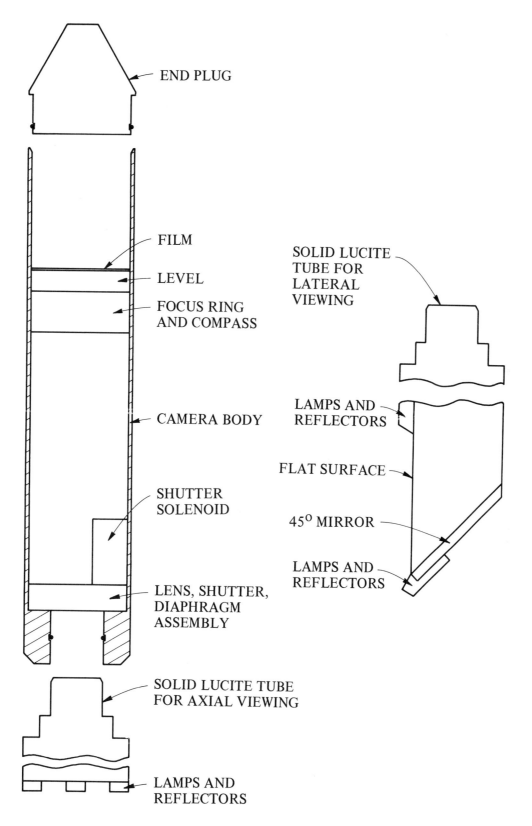

Figure 2. Borehole camera including tube for axial viewing at left; tube for lateral viewing at right. The outside diameter of the camera is 51 mm.

41

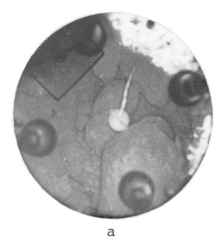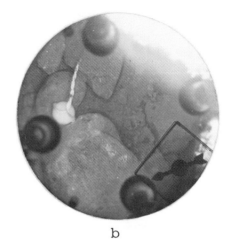

a   b

Figure 3. Photographs of the bottom of hole V1, 1969. (a) 7 September, 1245 hours, (b) 8 September, 1050 hours. The two bright areas in each photograph consist of ice attached to the borehole wall. Four lamp reflectors are seen, and a small nylon ball which rests on the bed. The portion of the bed seen has a diameter of about 87 mm.

Conclusions

The capability of drilling in dirty ice is probably necessary to reach the bed of a glacier in many instances. The cable tool and sand pump, used together with thermal drilling, are very suitable for this, at least if no large rocks have to be penetrated. The efficiency of the cable tool system described here needs some improvement, since in order to determine "typical" subglacial conditions it is necessary to drill many holes.

When the necessary drilling capability is at hand, borehole photography is a powerful method of studying subglacial conditions and motion. The turbidity problem has been partially solved by pumping, but some further work on the origin and the precipitation of the suspended material is needed. Until this is done the use of a more elaborate camera or television system probably is not warranted.

Our experience indicates that the necessary improvements can probably be made without sacrificing the low cost and portability of the equipment.

Acknowledgments

We are indebted to Henry Bell for engineering assistance. The work was done with the permission of the National Park Service, and the principal financial support was from the National Science Foundation. One of us (W.D.H.) acknowledges the support of State of Alaska Funds.

REFERENCES

Harrison, W.D., 1972, Temperature of a temperate glacier: *Journal of Glaciology,* v. 11, no. 61, pp. 15-29.

Harrison, W.D. and Barclay Kamb, 1973, Glacier borehole photography: *Journal of Glaciology,* v. 12, no. 64, pp. 129-137.

Johnson, E.E., 1966, *Ground water and wells: a reference book for the waterwell industry:* St. Paul, Edward E. Johnson, Inc.

Kamb, W.B. and R.L. Shreve, 1966, Results of a new method for measuring internal deformation in glaciers: *American Geophysical Union Transactions,* v. 47, no. 1, p. 190.

Shreve, R.L. and W.B. Kamb, 1964, Portable thermal core drill for temperate glaciers: *Journal of Glaciology,* v. 5, no. 37, pp. 113-117.

Shreve, R.L. and R.P. Sharp, 1970, Internal deformation and thermal anomalies in lower Blue Glacier, Mount Olympus, Washington, U.S.A.: *Journal of Glaciology,* v. 9, no. 55, pp. 65-86.

MECHANICAL PROPERTIES OF ANTARCTIC DEEP CORE ICE

A. Higashi and H. Shoji
Department of Applied Physics
Faculty of Engineering
Hokkaido University
Sapporo 060, Japan

ABSTRACT

Tensile tests were carried out with Antarctic deep core ice obtained at Byrd Station in 1968. Specimens for the tests were so prepared as to have their long axes parallel, perpendicular and inclined 45° with respect to the axis of core, or to the vertical direction of the ice sheet. Stress-strain relationships were recorded with different strain-rate and also at different temperatures.

The stress-strain curve generally exhibited a type of stress-saturation without any yield drop which was a characteristic feature in the case of easy glide in single crystals. This saturated value of the stress was considered as the maximum stress or the yield value. The relationship between the strain-rate $\dot{\epsilon}$ and the maximum stress $\sigma_{max.}$ was expressed as

$$\dot{\epsilon} = A \, \sigma_{max.}^{n}$$

The number of the exponent n was approximately 2, a little larger than that in the case of single crystals but smaller than those obtained for polycrystalline ice crystals hitherto obtained. The stress level which is designated by A in the equation above varied with the orientation of specimens, especially when the fabric diagram of specimens showed strong preferred orientations.

The experimental results are well interpreted by feasibility of the basal glides in polycrystalline aggregates with respect to the preferred orientations in ice specimens.

UNIVERSITY OF MINNESOTA ICE DRILL

Roger LeB. Hooke
Department of Geology and Geophysics
University of Minnesota
Minneapolis, Minnesota 55455

ABSTRACT

The University of Minnesota ice drill is an electrically-powered, portable, thermal drill designed for boring and coring to depths of a few hundred meters in polar glaciers. A new hot-point design is used in which a cylindrical tip makes an initial hole, and a parabolic section enlarges this hole to the desired diameter. Meltwater produced is diluted with ethylene glycol and left in the hole to counterbalance the hydrostatic pressure in the ice, and thus inhibit hole closure.

Introduction

In the spring of 1974 a drill for use in cold glacier ice was designed and built at the University of Minnesota. This drill (Figs. 1 and 2) consists of an electrical system including generator, variac, transformer, cable, and hot-point or core barrel; and an ethylene-glycol system including glycol reservoir, meter, pump, and tubing. Glycol is injected immediately above the hot-point to prevent refreezing of meltwater when drilling in cold ice.

The complete drill with 5-cm and 9-cm-diameter hot-points, 9-cm core barrel, and 5-kW generator can be built for approximately $7000 and weighs about 500 kg, excluding fuel and glycol. In field tests penetration rates were between 7.6 and 8.2 m/hr with the 5-cm hot-point and about 4.9 m/hr with the 9-cm hot-point.

Most components of the drill warrant only brief description, as there is nothing unique in their design. Two elements, however, are somewhat novel, and these will be treated in greater detail. One is a switch which signals the operator to let out more cable, and the other is the hot-point.

Electrical System

Power for the drill is provided by a 5-kW gasoline-powered Onan model 5CCK generator with 120- and 240-V outlets. Power from one of the 240-V outlets is run to a Superior Electric Co. model 236BU-2 powerstat, and thence through a step-up transformer which doubles the voltage (Fig. 3). With the use of the powerstat the voltage at the control-box outlet can be varied continuously from zero to approximately 560 V. High voltages are desirable to minimize power

Figure 1. Photograph of drill in operation, June 1974. The control box and cable reel are on the nearer sled, along with a box for tools and spare parts. The generator is on the further sled; a fuel drum and pump stand behind the generator.

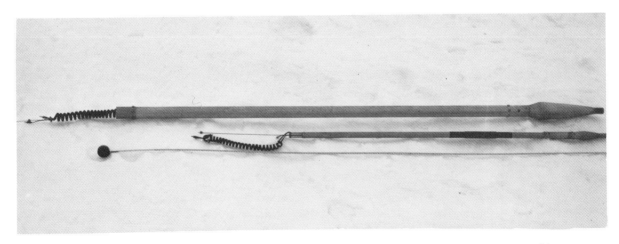

Figure 2. Hot-points and buoyancy sections. Coiled cable and switch wire are at top of buoyancy sections. Black area on buoyancy section for 5-cm hot-point is location of an emergency field repair job. Tape is 2 m long.

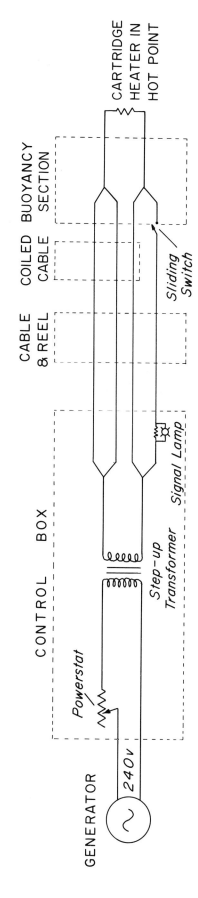

Figure 3. Schematic diagram of electrical system showing circuit for sliding switch.

49

losses in the cable. A flexible neoprene-jacketed type SO PWC cable with four No. 16 conductors is used to carry the power from the control box to the hand-operated cable reel, via a set of four slip rings, and thence down the hole to the hot-point. Allowable carrying capacity of this cable is 7 amperes per conductor.

The submerged part of the drill consists of a brass or copper hot-point or core barrel, and a buoyancy section (Fig. 2). The length of the buoyancy section is adjusted so that when submerged with the hot-point attached, it will assume a vertical attitude and thus keep the hole plumb (Aamot, 1968). To transfer power from the cable to the hot-point, three of the four conductors are connected to a short length of coiled 2- or 3-conductor cable (Figs. 2 and 3). One of these three leads is grounded to the buoyancy section and hence to the hot-point. The other two leads are joined and fed into the buoyancy section through an XSA-BC insulated, water-tight feed-thru, manufactured by Brantner & Associates, Inc., 3462 Hancock Avenue, San Diego, California. An insulated wire inside the buoyancy section, a second feed-thru at the lower end of the buoyancy section, and a third feed-thru on the hot-point (Fig. 4), with a wire between the latter two feed-thrus, completes this side of the circuit.

Switch

One of the four conductors in the main cable has not yet been accounted for. This conductor is connected to the upper end of a length of stiff, straight, springy wire, about 1 m long, which is clamped securely to the main cable. The lower end of this wire slides down into a 4-mm I.D. stainless-steel tube passing entirely through the buoyancy section from one end to the other. A cylindrical brass sleeve with rounded tip is slipped over the end of the wire and silver-soldered in place. This sleeve makes electrical contact with the tube and hence with the buoyancy section. When retracted from the stainless-steel tube, the sleeve enters a short thick-walled leucite tube and electrical contact is broken. The inside diameter of the leucite tube decreases at the top so the brass sleeve cannot be fully withdrawn. When the drill is being raised the weight of the hot-point is borne by the wire, the brass sleeve, and the leucite tube. Thus to provide additional strength at this point, the leucite tube is itself encased in a stainless-steel sleeve, which, however, does not make contact with the wire or brass sleeve.

When in operation, the cable is held fixed while the hot-point and buoyancy section penetrate into the ice. The coiled cable stretches, and the buoyancy section slides down the wire until the brass sleeve is retracted into the leucite tube. Current then ceases to flow in this conductor, but power still reaches the hot-point through the lead in the coiled cable which is grounded to the buoyancy section (Fig. 3).

As long as current flows through this switch, the voltage drop across a 1-ohm resistor lights a signal lamp on the control box (Fig. 3). When the brass sleeve is retracted into the leucite tube, however, the circuit is broken and the lamp goes out, at which time the alert operator lets out more cable.

Hot-Point

There appear to be two basic concepts in hot-point design. In one, used by the University of Washington (P.L. Taylor, written communication, 1974), a cylindrical cartridge heater is inserted into a hole in a solid copper cylinder. The curved surface and upper end of this cylinder are insulated to inhibit loss of heat radially and upward. Heat, therefore, is conducted downward to the lower end of the cylinder where melting occurs.

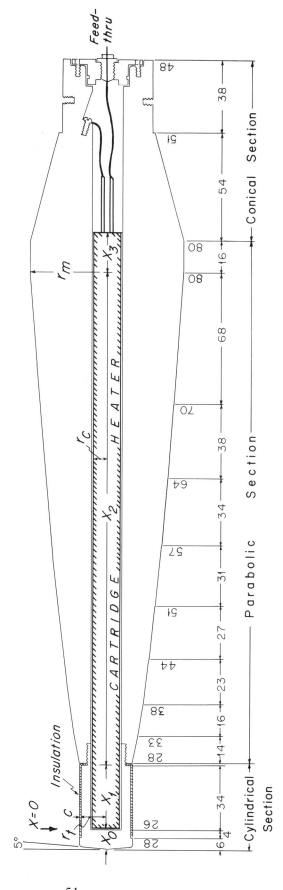

Figure 4. Cross section of 9-cm hot-point showing diameter of point at various positions, distances between these positions, insulation on cylindrical section, and definitions of variables used in calculations. All dimensions in millimeters.

In a second concept (Kasser, 1960) the hot-point is axially symmetric, say about the x-axis, and has a parabolic surface profile, $r \propto \sqrt{x}$, where r is the radius at a distance x from the tip. Heat is conducted radially outward to this surface, where melting occurs. The shape of the parabola is designed to satisfy the condition that each section of length Δx have just the melting capacity necessary to melt the ice encountered as the hot-point penetrates the ice at a uniform velocity. For a line heat source of length X located along the x-axis this condition is satisfied if r is given by·

$$r = \sqrt{\frac{x}{X}} r_m \qquad \qquad \text{Eq. (1)}$$

where r_m is the maximum diameter of the hot-point.

In hot-points which utilize a cylindrical cartridge heater as a heat source, the parabolic profile cannot be used at the tip because the radius of the tip must exceed that of the cartridge heater, and r, therefore, cannot decrease to zero at $x = 0$ (Eq. (1)). However, the parabolic design has two advantages over the cylindrical design. Firstly, in the parabolic design temperatures in the hot-point are substantially lower, thus reducing the possibility of failure of the heater. Secondly, construction is simpler because of the lower temperatures and because insulation is not required to prevent radial heat loss.

The present hot-point incorporates aspects of both designs. At the tip there is a cylindrical section in which radial heat loss is inhibited by insulation (Fig. 4). Heat is thus conducted downward and makes an initial hole. Above the cylindrical section there is a parabolic section in which heat is conducted radially outward and used to enlarge the hole. Finally, there is a conical section in which the diameter of the point is reduced to the diameter of the buoyancy section. It seemed advisable to make the buoyancy section somewhat smaller in diameter than the hole, lest some refreezing occur on the hole walls. As long as the buoyancy section remains free, heat lost through the conical section can be used to back out the hot-point.

The maximum temperature in the hot-point occurs immediately below the junction between the cylindrical and parabolic sections. This high temperature provides the temperature gradient down which heat is conducted to the tip. Heat loss back up into the parabolic section is inhibited by means of a rubber or teflon washer between the two sections (Fig. 4).

In order to select dimensions for the various sections such that the temperature at this junction is kept to a minimum, the temperature distribution in the cylindrical section was calculated under the assumption that radial temperature gradients are negligible. The temperature at the bottom of the cartridge heater (at $x = 0$ in Fig. 4) is given by:

$$T_0 = H \frac{x_1}{X} \frac{1}{\pi \eta r^2_t K_c} \sinh (\eta x_0) + T_b \qquad \qquad \text{Eq. (2)}$$

where H is the total heat generation by the cartridge heater in cal/sec, x_0 is the distance from the bottom of the cartridge heater to the blunt tip of the hot-point, x_1 is the length of the part of the cartridge heater which is in the cylindrical section, r_t is the radius of the cylindrical section, K_c is the thermal conductivity of the metal, T_b is the temperature at the ice-metal boundary, and $\eta = \sqrt{2 \dfrac{K_i}{c K_c r_t}}$ where c is the thickness of the insulation around the cylindrical section and K_i is the thermal conductivity of this insulation. The temperature at the junction between the cylindrical and parabolic sections is then obtained by numerically integrating the equation:

$$\frac{d^2 T}{dx^2} - \eta'^2 T + \xi = 0 \qquad \text{Eq. (3)}$$

over the interval $x = 0$ to $x = x_1$, using the boundary condition $T = T_0$ at $x = 0$. In this equation $\eta' = \sqrt{2 \dfrac{K_i}{c K_c} \dfrac{r_t}{(r_t{}^2 - r_c{}^2)}}$ where r_c is the radius of the cartridge heater, and $\xi = \dfrac{H}{\pi X K_c (r_t{}^2 - r_c{}^2)}$.

The shape of the parabolic section is now given by:

$$r = \sqrt{(r_t + c)^2 + \frac{x - x_1}{X} r_m{}^2} \qquad \text{Eq. (4)}$$

The term involving η' in Eq. (3) represents the effect of radial heat loss through the insulation. A similar term involving η appeared in the differential equation from which Eq. (2) was obtained by integration. This heat loss is:

$$H_l = H \frac{x_1}{X} (\cosh (\eta x_0) - 1) + \int_0^{x_1} K_1 \frac{T}{c} 2 \pi r_t \, dx \qquad \text{Eq. (5)}$$

The term involving ξ in Eq. (3) represents the effect of the heat source—the cartridge heater.

In order to transfer heat from one medium (say brass) to another medium which is in contact with the first (in this case water or ice), there must be a temperature difference, ΔT, between the two media. This temperature difference is proportional to the heat flux, q, thus:

$$\Delta T = \frac{q}{a}$$

where a is a coefficient of heat transfer.

Kasser (1960) notes that heat transfer from metal to ice is about three times as intensive as from metal to water. This is significant because heat is thus preferentially drawn to places where the hot-point is in contact with ice, and these are precisely the points where heat is required at any particular instant. However, the coefficient of heat transfer at a particular point may vary with time depending on whether the hot-point is in contact with water or with ice. Furthermore, if the hot-point is in contact with ice, the coefficient of heat transfer also depends upon the pressure at the contact point. Thus accurate calculation of ΔT is impossible, and the boundary temperature, T_b, cannot be determined theoretically.

In an attempt to measure T_b, five thermistors were embedded in the 9-cm hot-point. Leads to the thermistors were run down the outside of the hot-point, so this side of the point could not be in contact with ice during the experiment. Because heat transfer to water is less efficient, as noted, measured temperatures are expected to be somewhat higher than under actual operating conditions. In general, measured temperatures were 50 to 65°C higher than temperatures calculated on the assumption $T_b = 0$. Thus under normal operating conditions when the hot-point is largely in contact with ice it is estimated that T_b is approximately 20 to 30°C if H/X is 34 cal/cm sec.

53

Once the length and radius of the cylindrical section are selected, attention is directed to the conical section. Part of the cartridge heater extends into this section to provide heat for back out if necessary. During drilling this heat is transferred to the water and eventually either melts more ice, thus enlarging the hole, or is lost to the glacier. As the efficiency of this section is lower than that of the other two, its length should be kept relatively short. Regardless of its length, however, ice is melted by the heat generated in it, and the most efficient design is one which utilizes this heat to enlarge the hole to the desired diameter, rather than wasting the heat by making the hole larger than necessary. Thus the maximum diameter of the point is somewhat less than the design diameter of the hole.

On the basis of these several considerations, two hot-points were designed and built. The dimensions of the points are given in Table 1. The larger point was built first, tested in the laboratory, and then modified in accordance with the tests. The tests provided a basis for estimating the relative efficiencies of the three sections and these efficiencies were taken into consideration in the modifications. The small point was designed using these same relative efficiencies. However, in field tests of this point, the fall in water level in the hole as the hot-point was withdrawn suggested that the actual hole diameter was about 0.8 cm larger than design. This is interpreted as indicating that the efficiency of the cylindrical section was lower on the smaller point, probably because of the substantially lower pressure of the point against the ice.

It will be noted in Table 1 that the small point was designed for only 2000 W. Unfortunately the manufacturer of the cartridge heaters (Hotwatt, Inc., 128 Maple Street, Danvers, Massachusetts) could not supply higher-powered heaters that were sufficiently small in diameter to use in this small a point. This consideration, and the fact that the cylindrical section is more than one-quarter of the length of the point, suggest that the present design is not practical for hot-points less than 5 cm in diameter.

Calculated heat losses and maximum temperatures in the cylindrical section are also given in Table 1. The calculated value of H_l for the large point agreed well with estimates of the actual heat loss based on laboratory measurements. It also appeared that calculated values of T_{max} were at least qualitatively correct because rubber insulation was quite satisfactory on the cylindrical section of the large point, but burned up when used on the small point, indicating that the latter was substantially hotter, as predicted.

The overall efficiencies given in Table 1 are based on the ratio of the power required to melt ice which was initially at about -8°C, to the total power expended in the point. Line losses, which amounted to about 10 per cent of the total, are not included. The higher efficiency of the small point reflects its higher penetration rate, and hence the shorter time available for loss of heat by conduction to the ice surrounding the hole.

Core Barrel

A core barrel designed to be interchangeable with the 9-cm hot-point was built during the Spring of 1975. The barrel fits on the bottom of the buoyancy section, and takes a 50-cm core. We intend to take core every 5 or 10 m during drilling.

The barrel is modeled after that developed by the University of Washington (P.L. Taylor, written communication, 1974) and described elsewhere in this publication.

Table 1

Dimensions and Design Characteristics of Hot-Points

Point	Design hole diameter, cm	Actual hole diameter, cm	x_o, cm	x_1, cm	r_t, cm	c, cm	x_2, cm	x_3, cm	r_c, cm	r_m, cm	Design power, watts	Design current, amperes	H_l, cal/sec	Calculated $T max$, °C*
Small	4.5	~5.3	1.0	2.9	1.0	0.1	6.6	0.7	0.6	2.0	2000	9.1	9.8	143
Large	8.9	~9.0	1.0	3.4	1.3	0.1	25.1	1.9	0.8	4.0	4400	9.8	8.7	84

Point	Point Material	Insulating Material	Weight, kg	Estimated Efficiency, %	Penetration rate, m/hr
Small	Copper	Teflon tape	1.0	85	7.6 - 8.2
Large	Copper for cylindrical section Brass for rest	Rubber (bicycle tube)	8.7	75	4.9

*Assuming $T_b = 0$.

Glycol System

In polar glaciers with temperatures between 0 and about -15°C, hole closure can be a problem at depths of a few hundred meters if the hydrostatic pressure in the ice is not balanced by a comparable pressure in the borehole (Weertman, 1973). Therefore in the present system meltwater produced by the hot-point is diluted with ethylene glycol and left in the hole.

The glycol system consists of a reservoir, made from a 10-gal drum, whence glycol is drawn by a hand-operated piston pump and forced down the borehole through 0.96-cm I.D. plastic tubing. Between the reservoir and the pump there is a filter to remove foreign particles from the glycol, and, initially, a meter which recorded the total volume of glycol injected. The meter proved to be unsatisfactory as the glycol did not lubricate it properly. In its place a transparent tube and centimeter scale were installed on the side of the 10-gal drum to monitor glycol use.

The pump used is a piston pump designed by Kasser (1960) and capable of developing pressures of 5 to 10 kg/cm^2.

The plastic tubing used was black poly-flo tubing manufactured by Imperial Eastman Company, Chicago. Imperial Eastman also manufactures a variety of fittings which can be used with this tubing. The tubing is taped to the electrical cable, and both are handled simultaneously on the cable reel. The connection to the reel is made by means of a quick-release fitting on the hollow axle of the drum. This fitting doubles as a swivel coupling, although not actually designed as such. This cable system has proved to be awkward, however, due to differences in the coefficients of thermal expansion of the cable and tubing, and perhaps also to plastic stretching of the tubing. Alternatives are presently being investigated.

At the top of the buoyancy section a piece of latex surgical tubing, about 1 m long, is used to connect the plastic tubing to a 4-mm I.D. stainless-steel tube which passes entirely through the buoyancy section. This tube is parallel to the tube mentioned earlier for the switch. The surgical tubing is sufficiently elastic to stretch, along with the coiled electrical cable, as the hot-point penetrates into the ice. The glycol is thus injected immediately above the hot-point at the junction between the point and the buoyancy section.

When the drill is in operation, the reading on the glycol meter and the hole depth are recorded frequently, and the pumping rate is adjusted so that the concentration of glycol in the hole is just sufficient to prevent freezing at prevailing ice temperatures. The volume of glycol injected per meter of hole drilled will obviously depend upon hole diameter and ice temperature. The pump capacity should be such that this volume can be injected by an operator pumping at a leisurely rate 10 to 25 per cent of the time.

Concluding Statement

The drill was used on the Barnes Ice Cap, Baffin Island, N.W.T., in June and July 1974 and June 1975. Three 5-cm-diameter holes were drilled to the bed of the glacier at points near the margin where the ice was 100, 110, and 132 m thick respectively. These holes were cased with 3-cm O.D. aluminum casing and filled with diesel fuel. Temperature measurements were obtained in the two shorter holes.

A fourth hole was drilled to a depth of 43 m with the 9-cm hot-point in 1974. Penetration ceased at this depth due to an accumulation of sediment in the bottom of the hole. In 1975 a fifth hole was drilled near the fourth. Dirty ice was again encountered at a depth of 41 m, but with the use of the core barrel built in 1975 we were able to penetrate an additional 11.5 m of this dirty ice before drilling rates became unreasonably low (<0.1 m/hr). Both temperature and inclinometer measurements were obtained in this hole, and fabric studies have been completed on cores from it.

On the basis of our experience during the 1974 field season, there is no reason to believe that the drill will not be suitable for drilling to the design depth of 450 m, though considerable patience may be required for the deeper holes.

Acknowledgments

I would like to thank Lyle Hansen for providing the impetus for development of this drill, Philip Taylor for sharing with me his ideas on drill design, the National Science Foundation (Grants GA-19310 and GA-42728) for financial support, and the Glaciology Division, Department of Environment, Canada, for logistical support in the field.

REFERENCES

Aamot, H.W.C., 1968, A buoyancy-stabilized hot-point drill for glacier studies: *Journal of Glaciology,* v. 7, no. 51, pp. 493-498.

Kasser, P., 1960, Ein Leichter thermischer eisbohrer als hilfsgerät zur installation von ablationsstagen auf gletschern: *Geofisica Pura e Applicata,* Milano, Bd45, pp. 97-114.

Weertman, J., 1973, Anticipated closure rates for a proposed drill hole, Ross Ice Shelf, Antarctica: U.S. Army CRREL Special Report 190.

NEAR-SURFACE SNOW SAMPLING DEVICES*

S.J. Johnsen
Geophysical Isotope Laboratory
University of Copenhagen
Denmark

ABSTRACT

Firn cores obtained by presently available coring augers are normally not continuous in the upper 2-3 m. In stable isotope studies it is very important to obtain continuous samples from the surface down, which therefore have to be collected from a pit wall. Two kinds of sidewall samplers have been designed and tested that make collection of samples to a depth of 2 m or more from the surface an easy and fast operation. A short description of these two sampling devices is given. It is hoped that such sampling devices will be used by all future field parties in order to obtain surface samples for stable-isotope studies.

Introduction

In recent years a new glaciological parameter has become extremely important in modern glaciological studies. This parameter is the 3-dimensional distribution of stable-isotope ratios O^{18}/O^{16}, or D/H (measured in permille deviation from a standard, i.e. δ values), in glaciers and other large bodies of ice.

As discussed in Dansgaard *et al.* (1973), knowledge of this parameter can reveal information on the past history of glaciers. In particular, climatic changes occurring in the past are recorded in the stable-isotope ratios. This very important information is obtained from ice or firn cores where corrections usually have to be made due to the flow of ice and the changes in the isotopic ratios in surface snow upslope from the drill site.

Thus in most cases information about the geographic distribution of surface snow δ-values is required for interpreting δ-profiles from ice cores. Unfortunately, existing data on the distribution of surface δ-values is sporadic and often not reliable because the collection of samples has frequently been made under non-ideal conditions. A device for obtaining such samples under controlled conditions is therefore urgently required.

---

*Paper was presented by R.H. Thomas.

In order to facilitate sampling of surface snow for stable-isotope determination two kinds of samplers constructed at the present laboratory have been tested in both Greenland and Antarctica. Both samplers work in an already drilled hole, and there is no need to dig special pits.

Type 1 Sampler

This device is shown in Fig. 1. It is made from a perspex cylinder which attaches to a SIPRE auger drilling rod. A specially designed cutter on the side of the cylinder cuts the sample out of the wall of the drill hole as the sampler is moved up the hole. The sample then falls via a hole in the cylinder wall into a premarked sample bag, which is fixed to the bottom of the device. The samples obtained by this device are thus mean samples which are truly representative of the snow cover at the drill site. Samples can be taken at any depth in the drill hole.

The overall length of the sampler is 20 cm, it can go into an 11-cm or larger diameter hole, and it weighs 880 g. The cross section of the triangular-shaped sample is 3.5 cm^2; thus a 1-m-long sample represents 350 cm^3 of snow corresponding to approximately 100 g of water.

Experience shows that sampling is extremely easy and fast with this device. To drill a 2-m hole with a SIPRE auger and to sample the hole takes less than 15 min.

Field parties can thus collect several samples during a traverse without any significant burden to the field program.

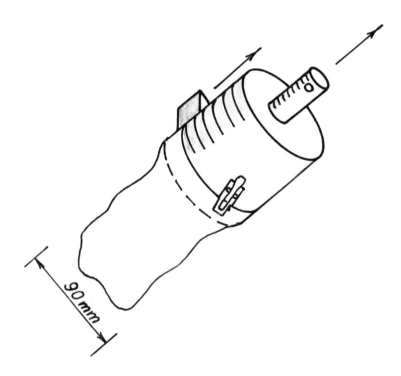

Figure 1. Type 1 Side-wall sampler.

Type 2 Sampler

Unlike the type 1 sampler, this type is made for obtaining more detailed profiles from the hole, and is therefore a much heavier and more complicated device (see Fig. 2). It consists of an 11-cm-diameter, 210-cm-long perspex tube with a long slit on one side of it. Inside the tube is mounted, on bearings close to the wall, a 6-cm-diameter, 200-cm-long half-cylinder "sawblade." It can be turned along its center, in the bearings mentioned, out through the long slit by turning the handle on top, at the same time the entire sawblade can be moved up and down 10 cm. By doing so with the sampler fixed in a borehole a half core of 6 cm diameter is then cut out of the wall of the hole. A sharp half-circular blade is at the same time turned into the wall right below the half cylinder sawblade, thus supporting the half core at the bottom.

The entire sampler is then taken out of the hole along with the 2-m half core which is well protected between the wall of the perspex tube and sawblade. Stratigraphic features can then be recovered from the half core which then is cut into an appropriate number of increments. Experience shows that even very soft and coarse layers of snow are undisturbed in the 2-m-long half-core sample.

In very low accumulation areas the upper 2 m of firn are often very soft and coarse grained, without any hard layers. Using the sampler in such areas has not given satisfactory results. Changes are now being made on the sampler that are hoped to improve greatly the performance of the device in soft snow.

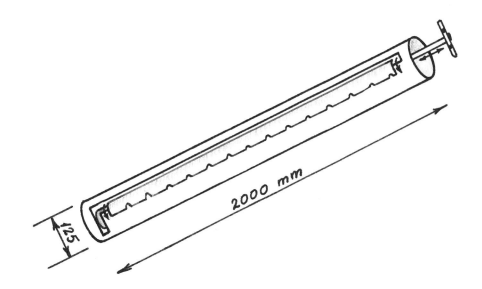

Figure 2. Type 2 Side-wall sampler.

REFERENCE

Dansgaard, W., S.J. Johnsen, H.B. Clausen, and N. Gundestrup, 1973, Stable isotope glaciology: *Meddelelser om Grønland,* v. 197, no. 2, 53 p.

ICE SHEET DRILLING BY SOVIET ANTARCTIC EXPEDITIONS

Ye. S. Korotkevich
Arctic and Antarctic Institute
and
B.B. Kudryashov
Leningrad Mining Institute
Leningrad, U.S.S.R.

ABSTRACT

Scientists of the Leningrad Mining Institute and of the Arctic and Antarctic Research Institute have developed fundamentals of the theory of the process of drilling by melting of open holes. A series of electrothermal core drills used with a cable has been manufactured. Two drilling rigs have also been developed and manufactured: a stationary one for deep drilling, and a mobile one for drilling to 150 m.

The deepest borehole at Vostok Station is 952 m. Core recovery was 99 per cent. The core diameter is 125 mm, and the borehole diameter is 180 mm.

Geophysical and geochemical studies in the boreholes have been made together with observations of ice-sheet dynamics and ice rheology.

The Arctic and Antarctic Research Institute has developed and successfully used lightweight electrothermal core drills for sinking boreholes in warm mountain glaciers without pumping out the water. In addition, thermal drills for drilling without core removal in warm and cold glaciers have been developed. Depths of boreholes in warm glaciers (with and without core removal) and in cold glaciers (without core removal) are 200 m and 80 m, respectively.

At the present, an electromechanical core drill is being tested in laboratory and field conditions, thermal core drills for a non-freezing fluid-filled hole are being developed, effective fluids are being selected, a system of removing ice chips and meltwater from the borehole is being developed, and the equipment for drilling to 4000 m is being designed.

Preparation is underway for drilling a deep borehole at Vostok Station in the 1977-1978 season.

Electrothermal drilling devices for coring and non-coring melt penetration of holes are used for ice-sheet studies in Soviet Antarctic Expeditions (Barkov, 1960; Barkov et al., 1973; Bobin and Fisenko, 1974; Korotkevich, 1973; Kudryashov and Yakovlev, 1973; Kudryashov et al., 1973; Morev, 1972a, 1972b; Sekurov, 1967).

A method for calculation is developed which enables us to select optimum variants of construction elements and regimes of drilling by melt penetration. On the basis of the theory of moving sources of heat (Rosenthal, 1946) and the theory of heat conductivity of the bodies with dispersed heat sources by means of conjugation of one-dimensional temperature fields, a stationary process of melt penetration drilling is analytically outlined, with the heat exchange between the heater and the ambient liquid and gaseous medium being taken into account.

A general formula of the rate of melt penetration drilling of open holes is obtained (Kudryashov and Fisenko, 1973).

$$V = \frac{2ka\lambda_1 m\,(1 - nL^2)\,L}{a\rho\gamma - (T + \sigma H)\,\lambda_2}$$

where

$$m = \frac{1}{2\lambda_1}\left[\frac{1096N}{(D_2^2 - D_1^2)\,L} + \alpha\pi\,(D_1 + D_2)t\right]$$

Eq. (1)

$$n = \frac{\alpha\pi}{2\lambda_1}(D_1 + D_2) \quad ; \quad k = \frac{F_1}{F_2}$$

D_1 and D_2 — inner and outer diameters of the annular heater

F_1 and F_2 — areas of the hole bottom and heater cross section

L — height of the heater

T — mean annual temperature of the ice surface

t — mean temperature of the ambient liquid and gaseous media

N — electric capacity of the heater

λ_1 and λ_2 — heat conductivity of the heater and ice materials

a — temperature conductivity of the ice

ρ — melting heat of the ice

γ — ice density

α — coefficient of heat exchange between the heater and ambient liquid and gaseous medium

σ — ice temperature gradient

H — depth of the hole

Eq. (1) considers major factors that determine the process of melt penetration drilling. Its analysis makes it possible to assess the influence of these factors upon the effectiveness of

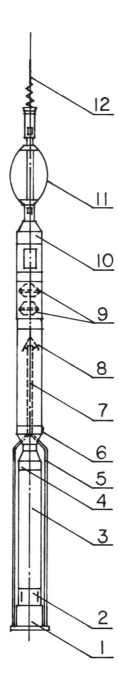

Figure 1. Schematic diagram of the electrothermal drill TELGA (see text for explanation of numbered areas).

the process. The calculated values of the rate of melt penetration drilling have proved to be close to the observed ones (Barkov *et al.,* 1973; Bobin and Fisenko, 1974; Kudryashov *et al.,* 1973; Kudryashov and Yakovlev, 1973).

The electrothermal drill TELGA used in the Soviet Antarctic Expedition is shown schematically in Fig. 1. The thermal drill has an annular heater consisting of a copper body which contains a nichrome heater element in a ceramic insulation. Air insulation between the wall of the hole and the heater is provided by ring-shaped protrusions on the bit bottom. Above the heater there is a core catcher (2). Along the core barrel (3) there are drainage pipes (5), the lower ends of which are close to the bit bottom and make it possible to control water ejection.

The upper ends of the pipes (5) are fixed on the assembly adapter (4) which connects them with the inner pipe (7) of the water tank (8). The adapter (4) has a valve (6) to drain water after the drill is raised. In the upper part of the tank (8) there is a vacuum-pump section with a pair of two-stage turbo-pumps (9) joined in sequence. The armor of the cable-wire (12) is fixed to the upper adapter (10). Above the adapter (10) there is a spring centering device (11).

The outer and inner diameters of the annular heater are 178 mm and 130 mm, respectively. The electric capacity of the heater is 3.5 kW, the length of the core barrel is 2 m, and the total length of the drill is 7.5 m.

The TELGA electrothermal drill operates as follows. After lowering the drill in the hole the heater (1) and vacuum pumps (9) are switched on. Due to vacuum in the tank (8) the water that forms on the bottom is ejected by the air flow and transported through the pipes (5) and (7) into the tank (8).

After the core barrel (3) is filled, the power supply is switched off, the hoist raises the drill from the bottom of the hole and the core catcher (2) automatically grips and retains a core. At the surface the core is retrieved, the water drains from the tank and the drill is prepared for the next run.

A test drilling was made at the 50th kilometer from Mirny, Antarctica, in October-November 1969. During 36 days a hole 250 m deep was completed and a core was recovered (Kudryashov *et al.,* 1973).

In 1970, with the help of the TELGA electrothermal drill, a deep hole was started at Vostok Station. In four months a depth of 507 m was achieved and a core was recovered (Barkov *et al.,* 1973; Kudryashov, *et al.,* 1973). A schematic diagram of the drilling equipment is given in Fig. 2, where

    1— electrical switchboard

    2— control board

    3— window aperture

    4— oil furnace

    5— auxiliary hoist

    6— rig

    7— tower rings

    8— installation of the drill over the hole

9— supporting framework

10— bench and instruments

11— fire prevention capacity

12— steel sledge

13— hoist motor

14— two-speed gear reducer

15— main hoist

16— cable wire

17— generator-motor system

18— lathe

19— collapsible balance-block

20— drill in the hole

21— conductor

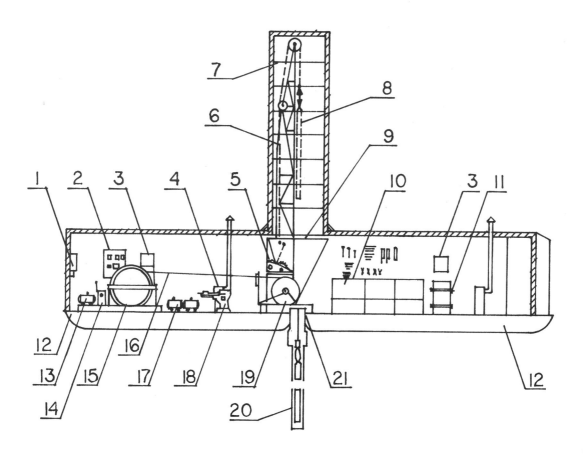

Figure 2. Schematic diagram of the electrothermal drilling equipment at Vostok Station (see text for explanation of numbered areas).

In 1971 complications occurred in the course of drilling due to lack of reliability of the control system for the performance of the drill.

In May 1972 the depth of the hole was 952 m, and a core was recovered.

In 1973 an attempt to continue drilling by melt penetration of a deeper open hole failed because of hole closure. As a result of this failure a part of the hole was left uncompleted and a new hole was started at a depth of 350 m. At present there is an open hole 905 m deep at Vostok Station. In this hole observations on temperature, deviation from the vertical, and the rate of deformation of the wall of the hole are being made.

For field conditions a mobile drilling unit, installed on a sledge, with a mobile version of the TELGA drill has been used (Bobin and Fisenko, 1974). During research traverses along the route Mirny-Pionerskaya three core holes to 70 m in depth were completed. The total depth of the holes completed by the TELGA drills was about 3000 m, with a depth of 952 m as a maximum.

For non-coring drilling of shallow holes lightweight electrothermal drills have been developed. These drills have conical heaters 40 and 80 mm in diameter (Morev, 1972b). During drilling in cold snow, water percolates into the wall of the hole, and during drilling in warm mountain ice with ice temperature about 0° C, the hole remains filled with water.

A coring electrothermal drill with an annular heater (Morev, 1972b) was also tested in mountain ice.

The conical heater (Morev, 1972b) is shown in Fig. 3. Its main feature is as follows: part of the heat from an electric heater coil (1), wound on a copper core (2), is transferred to the heater point; the remaining part, in a quantity needed to expand the area of the bottom of the hole to a given extent, is transferred to the ice through the outer wall of the heater (3). The distribution of heat over the surface of the heater depends on the thickness of the copper core and side insulation (4). The heater is designed to operate in a borehole filled with water or non-freezing liquid. The heater connections (5) are sealed with rubber (6).

Core bits are of the same construction, but the copper core and the walls are ring-shaped. The coring electrothermal drill tested in warm ice has an annular heater, a core barrel, core catchers, a centering device and a cable joint. It was lowered in the hole on a single-core cable 12 mm in diameter. The steel armor of the cable was used as a second conductor. The total depth of the holes completed by this drill was about 4000 m; a maximum depth was 200 m. The core drill ran about 250 m, with a maximum depth of 120 m.

At the present, surface drilling equipment and cable-suspended electrothermal and electro-mechanical drills for drilling 4000-m-deep holes filled with a non-freezing liquid are being developed.

At the same time apparatus and techniques for glaciological, geophysical and thermophysical research, as well as instruments for recovering sterile ice samples for microbiological studies are being developed.

It is being planned to begin the development of a device for recovering CO_2 samples from the ice for radiocarbon dating.

Table 1

Specifications of Electrical Heater Bits

In Electrothermal Drills

| | | TYPES OF HEATER BITS | | | Flat working surface (TELGA) |
| | | Conical working surface | | | |
		continuous		annular	annular
Diameter, mm	heater hole core	48 48-50 -	80 82-85 -	112/88 116-120 82-85	178/130 180-185 120-125
Ice temperature, $^\circ$C		0	-19 to -28	0	-28 to -57
Power, kW		1-2	3-4	1.5-2.2	3.5
Voltage, V		150	200	160	250
Drilling rate, m/hr		7-9	7-10	3-4.5	1.5-2

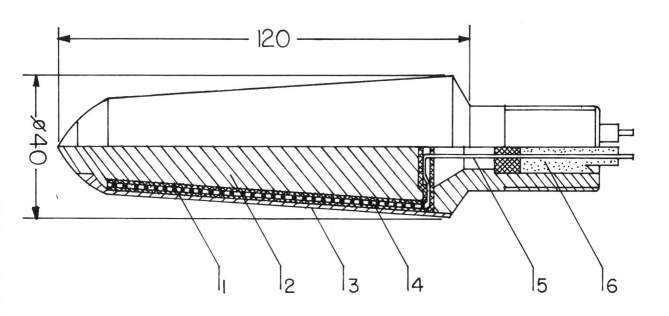

Figure 3. Schematic diagram of a conical electrical heater bit (see text for explanation of numbered areas).

REFERENCES

Barkov, N.I., 1960, Electrothermal drill for drilling holes in the ice: *Bulletin of Inventions,* no. 8, license no. 127629.

Barkov, N.I., N.E. Bobin and G.K. Stepanov, 1973, Drilling of a bore hole in the ice sheet of Antarctica at Vostok Station in 1970: *Soviet Antarctic Expedition Information Bulletin,* no. 85, pp. 22-28.

Bobin, N.E. and V.F. Fisenko, 1974, An experience of thermal core drilling of boreholes in field conditions: *Soviet Antarctic Expedition Information Bulletin,* no. 88, pp. 74-76.

Korotkevich, Ye.S., 1973, Discussions on the problems of the study of the Antarctic ice sheet at the XIIth SCAR meeting and at the meeting of the IAGP Council: *Materials of Glaciological Research. Chronicle. Discussions,* Issue 21, pp. 44-50.

Kudryashov, B.B. and V.F. Fisenko, 1973, Theory of drilling by melt penetration in snow-firn and ice in Antarctica: *Antarctica Commission Reports,* Issue 12, pp. 153-158.

Kudryashov, B.B. and A.M. Yakovlev, 1973, A new technology of drilling holes in frozen bedrock: Leningrad, "Nedra," 168 p.

Kudryashov, B.B., V.F. Fisenko, G.K. Stepanov and N.E. Bobin, 1973, An experience of drilling in the ice sheet of Antarctica: *Antarctica Commission Reports,* Issue 12, pp. 145-152.

Morev, V.A., 1972a, On the efficiency and economics of electrothermal drills in the drilling of inland ice: *Soviet Antarctic Expedition Transactions,* v. 55, pp. 158-165.

Morev, V.A., 1972b, A device for electrothermal core drilling: *Bulletin of Inventions and Discoveries,* no. 27, license no. 350945.

Rosenthal, D., 1946, The theory of moving sources of heat and its application to metal treatments: *American Society of Mechanical Engineers, Transactions,* v. 68, pp. 849-866.

Sekurov, A.V., 1967, Peculiarities of the development of an electrothermal unit for drilling in ice and results of its test at Mirny in 1965-66: *Soviet Antarctic Expedition Information Bulletin,* no. 60, pp. 59-62.

THE POLAR ICE-CORE STORAGE FACILITY

AT USA CRREL*

Chester C. Langway, Jr.†
U.S. Army Cold Regions Research and Engineering Laboratory
Hanover, New Hampshire 03755

Since the inception of the U.S. polar ice-core drilling program, the U.S. Army Cold Regions Research and Engineering Laboratory (USA CRREL) has been responsible for the central storage and curatorial activities of the ice cores recovered in the Office of Polar Programs/National Science Foundation (OPP/NSF) Arctic and Antarctic research programs (Table 1).

The main purpose of the central ice-core storage facility is to handle, process, catalog and distribute the ice cores drilled in the polar regions (Lange, 1973; Langway and Hansen, 1970; Langway *et al.,* 1970; Ueda and Garfield, 1968, 1969a, 1969b) to OPP-approved recipients for glaciological research. Under an agreement with OPP, the ice cores are stored at USA CRREL and in a commercial freezer facility; a technician handles and catalogs them. A core data bank is maintained for retrieval and information exchange, and starting with the Dye 3 ice core, is being computerized.

Between 1956 and 1968, USA CRREL had within its own physical plant the cold room capacity to store the ice cores. With the advent of core drilling through both the Greenland and Antarctic ice sheets in 1966 and 1968, respectively, USA CRREL's available storage capacity was exceeded and a nearby commercial outlet was located.

Within the present USA CRREL cold room complex a 15 feet (4.6 m) by 25 feet (7.6 m) room, 16 feet (4.9 m) high, maintained at -34 ±2°C is reserved solely for the storage of ice cores and other surface samples (Room 161, Fig. 1). Approximately 400 5-foot (1.5 m) core tubes from selected depths over the profiles of the various ice cores listed in Table 1 are stored in Room 161. These cores are used in USA CRREL's in-house ice-core analysis program as well as for out-of-house needs. Adjacent to and having the same dimensions as the storage room (Room 160, Fig. 2) is the cold laboratory (-10 ±2°C) used for core processing, chemical cleaning of cores, various physical property studies (within a dust-free hood), and photography.

---

*This paper is also printed in the *Antarctic Journal of the United States,* v. 9, no. 6, 1974.

†Present address: Department of Geological Sciences, State University of New York at Buffalo, Amherst, New York 14226.

Table 1

Arctic and Antarctic Ice-core Inventory

Year Drilled	Location	Drilling Method	Depth of Drilling, m
		PRE-IGY	
1956	Site 2, Greenland	Rotary	305
1957	Site 2, Greenland	Rotary	411
		IGY	
1958	Byrd Station, Antarctica	Rotary	308
1959	Little America V, Antarctica	Rotary	256
		POST-IGY	
1961	Camp Century, Greenland	Thermal	185
1962	Camp Century, Greenland	Thermal	235
1966	Camp Century, Greenland	Electromechanical	1375
1968	Byrd Station, Antarctica	Electromechanical	2164
		GISP	
1971	Dye 3, Greenland	Thermal	372
1973	Station Milcent, Greenland	Thermal	398
1974	Station Crete, Greenland	Thermal	405

The bulk of the ice cores are stored at a commercial storage facility in Littleton, New Hampshire, 60 miles (96 km) north of USA CRREL. A total of 820 square feet (76.2 m^2) of floor area is rented consisting of three racks 16 feet (4.9 m) high (Fig. 3). Each rack is 40 feet (12.2 m) long and divided into 13 units in length and 5 units in height. Each unit holds between 17 and 22 core tubes which vary in diameter to accommodate the 4 inch (10.2 cm) to 4-7/8 inch (12.4 cm) diameter ice cores. As many as 20,000 core tubes can be stored here. The ice cores are sheathed in polyethylene tubing and contained in specially constructed near-vacuum-tight aluminum-foiled tubes in the field. The commercial cold-storage facility is kept at -24 ±2°C using ammonia as the refrigerant. It has a complete back-up system and a separate power plant. The entire building was constructed and insulated to remain below freezing for 8 days after a power failure. In addition, sufficient refrigerated trailers are always available in the event of a catastrophe.

Figure 1. Cold room 161 at USA CRREL used exclusively for ice core and other polar sample storage. Temperature of storage is -34 ±2°C.

Figure 2. Cold room 160 at USA CRREL used for ice-core processing and study. Temperature of laboratory is -10 ±2°C.

Numerous samples of Arctic and Antarctic cores have been supplied to investigators around the world, and results of studies have been published in various trade journals (see bibliography by Langway and Gow, 1968). These investigations encompass all three components of the material making up an ice core: entrapped air, the ice itself and foreign matter—both particulate and dissolved. Many analytical techniques for analysis of the recovered cores have been developed, but recognizing the extreme value of the cores, as well as the limitations of any particular laboratory in total analytical technology, an integrated and cooperative analysis program has been built in which a large number of investigators have participated. Main participation is presently being shared by three interrelated groups, each responsible for a particular analytical area:

USA CRREL	Physical and chemical analyses
University of Copenhagen	Stable isotope and particle analyses
University of Bern	Radioactive isotope and gas analyses

In addition to the above, the Ohio State University Institute of Polar Studies is developing a particle analysis laboratory under OPP sponsorship. Various other investigators have and are performing specific studies on the different ice cores using grants from other federal, state, educational and private institutions (Langway, 1974).

Figure 3. Commercial ice-core facility north of USA CRREL. On the right are Camp Century, Greenland, ice cores. In the center is fork-lift truck emplacing a crated shipment of ice cores from Byrd Station, Antarctica. Core racks are 40 feet (12.2 m) long, 5 feet (1.5 m) wide and 16 feet (4.9 m) high. Temperature of storage is -24 ±2°C.

Concurrent with development of glacier drilling technology, USA CRREL and the University of Copenhagen have developed a core stratigraphy and logging routine and surface pit/ice-core correlation system which assures accurate and consistent recording of stratigraphic features providing a firm chronological, geochemical and isotopic datum for all studies.

Samples of the ice cores are made available to any interested scientist that is funded through OPP for this purpose or has an on-going research effort that would benefit from obtaining polar ice core samples. To obtain samples, in the first case, submit proposal through OPP. In the latter case, write to the author.

The ice-core facility is supported by National Science Foundation contract CA-23.

REFERENCES

Lange, G.R., 1973, Deep rotary core drilling in ice: U.S. Army CRREL Technical Report 94.

Langway, C.C., Jr., 1974, Outline of investigations on the deep ice cores from Greenland and Antarctica: U.S. Army CRREL Technical Note Series.

Langway, C.C., Jr. and A.J. Gow, 1968, Selected bibliography on the USA CRREL deep core drilling in ice and ice core analysis program: U.S. Army CRREL Technical Note Series.

Langway, C.C., Jr. and B.L. Hansen, 1970, Drilling through the ice cap: Probing climate for a thousand centuries: *Bulletin of the Atomic Scientists,* v. 26, no. 10, pp. 62-66.

Langway, C.C., Jr., A.J. Gow and B.L. Hansen, 1971, Deep drilling into polar ice sheets for continuous cores: in *Research in the Antarctic,* L.O. Quam, Editor, Washington, D.C., American Association for the Advancement of Science, Publication No. 93, pp. 351-365.

Ueda, H.T. and D.E. Garfield, 1968, Drilling through the Greenland Ice Sheet: U.S. Army CRREL Special Report 126.

Ueda, H.T. and D.E. Garfield, 1969a, The USA CRREL drill for thermal coring in ice: *Journal of Glaciology,* v. 8, no. 53, pp. 311-314.

Ueda, H.T. and D.E. Garfield, 1969b, Core drilling through the Antarctic Ice Sheet: U.S. Army CRREL Technical Report 231.

GENERAL CONSIDERATIONS FOR DRILL SYSTEM DESIGN

Malcolm Mellor and Paul V. Sellmann
U.S. Army Cold Regions Research and Engineering Laboratory
Hanover, New Hampshire 03755

ABSTRACT

Drilling systems are discussed in general terms, component functions common to all systems are identified, and a simple classification is drawn up in order to outline relations between penetration, material removal, hole wall support, and ground conditions. Energy and power requirements for penetration of ice and frozen ground are analyzed for both mechanical and thermal processes. Power requirements for removal of material and for hoisting of drill strings are considered, and total power requirements for complete systems are assessed. Performance data for drilling systems working in ice and frozen ground are reviewed, and results are analyzed to obtain specific energy values. Specific energy data are assembled for drag-bit cutting, normal impact and indentation, liquid jet attack, and thermal penetration. Torque and axial force capabilities of typical rotary drilling systems are reviewed and analyzed. The overall intent is to provide data and quantitative guidance that can lead to systematic design procedures for drilling systems for cold regions.

INTRODUCTION

Drilling involves an enormous range of highly specialized processes, products, and technologies, making it difficult to assimilate all the information required for solution of particular drilling problems. This difficulty is very pronounced in the case of problems that involve frozen ground and massive ice, since existing drilling systems are likely to require modification to meet the special ground conditions. It is therefore desirable to consider the basic elements of drilling systems that are often obscured by the technicalities and complications of practical products and processes.

In this short review, a scheme for classification and analysis of drilling systems is outlined as a preliminary step. The intention is to illustrate a broad systematic approach without attempting to cover each aspect of drilling in detail.

BASIC ELEMENTS OF DRILLING SYSTEMS

Virtually all practical drilling systems embrace three basic functions:

1.   Penetration of the ground material

2.   Removal of the surplus material

3.   Stabilization of the hole wall.

Each of these factors can be dealt with in a variety of ways, leading to a very large number of potential combinations for complete systems. However, the number of available combinations is reduced somewhat by the need for compatibility between individual elements in a practical drilling system.

Figure 1 outlines the main elements of practical drilling systems and indicates some compatibility links between individual elements. It does not embrace novel experimental drilling concepts such as hypervelocity water jets or electromagnetic devices, although such things could be added to the scheme.

Penetration

In most conventional drilling systems, penetration is accomplished by one of two methods: (a) direct mechanical attack, or (b) thermal attack.

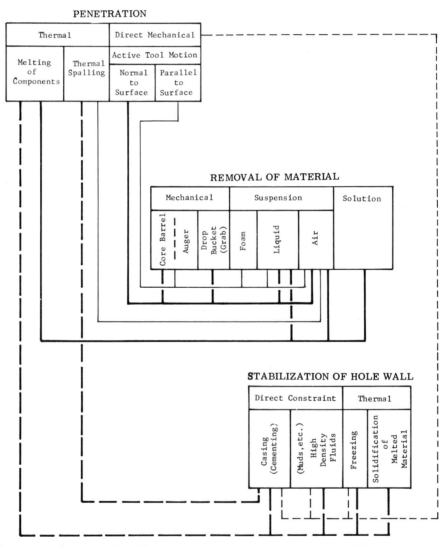

Figure 1. Elements of practical drilling systems and suggested compatibility links.

Direct mechanical processes can be broadly subdivided according to the working motion of the bit or cutting tool relative to the advancing surface. Motion is usually either parallel or normal to the advancing surface. Percussive bits and roller bits are examples of tools in which the cutting or chipping element moves normal to the advancing surface during the active stroke. In these cases of normal motion, the resultant force on the active component is also very nearly normal to the advancing surface. Drag bits and diamond bits are examples of tools in which the cutting element moves parallel to the advancing surface. However, the resultant force on the cutter tip of the parallel motion tool is not parallel to the surface, since a substantial normal component of force is usually involved.

Many considerations enter into the selection of a mechanical process, but the choice is heavily dependent on the properties of the material to be cut, particularly the strength, ductility and abrasiveness. Figure 2 gives a rough indication of the range of applicability for various types of bits and drilling systems.

Thermal penetration methods usually depend upon either (i) complete or partial melting of one or more components of the ground material, or (ii) thermal spalling in suitable materials. Melting methods have been widely used in ice and frozen soils; representative devices include electrically-heated thermal corers and probes for ice, and steam-point drills for ice and ice-rich mineral soils. Similar methods could be used in other materials with low melting point, e.g. sulfur. More novel melting devices are being studied experimentally for drilling and tunnel boring in hard rocks generally; these employ high temperature heating (up to about 2000°K) that is capable of melting and fusing silicates. Thermal spalling depends on development of large strains and high strain rates by rapid heating or cooling. The presence of strain discontinuities is also important. Certain types of rocks, known as "spallable rocks" (usually crystalline rocks with constituent minerals that may have widely differing expansion coefficients) are well suited to thermal spalling under the action of flame jets, plasma arcs, lasers, etc.

Jet penetration methods, which are still in the experimental stage of development, might be regarded as a special form of direct mechanical attack, although there may be some tenuous relations to thermal principles. Explosive shaped charges, in which interacting shock fronts form a jet and entrain metal particles, have long been used to punch shallow holes, but they have not been used for deep drilling (they have been considered for tunneling). Streams of free solid projectiles, which are basically similar in function to percussive tools, have been proposed for tunneling, but not for deep drilling. However, liquid jet drilling, using either a pure liquid or a liquid containing solid particles, is under active development. No jet drills have yet been built for use in ice or frozen ground, but basic experiments with jet pressures up to 100,000 $lbf/in.^2$ (690 MN/m^2)*

Note on Units. The primary units in this paper are American Units, since much of the relevant technology and much of the source material involves numbers that are rounded in this system. SI equivalents are given in parentheses as far as possible, and to cover those instances where dual units are not practicable, the following conversion factors are offered:

American Unit	Multiply by	To obtain SI Unit
in.	25.4	mm
ft	0.3048	m
ft/sec	0.3048	m/sec
ft/min	5.08	mm/sec
lbf	4.448	N
$lbf/in.^2$	6.895×10^3	N/m^2
or $in.-lbf/in.^3$	6.895×10^3	J/m^3
ft/lbf	1.356	J
hp	0.7457	kW

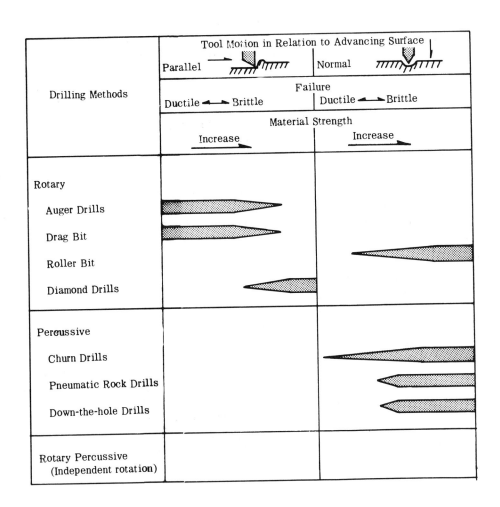

Figure 2. Range of applicability of various types of drilling methods in relation to tool motion and material properties.

have been carried out on ice and frozen soils, and rotating nozzle systems applicable to drilling have been developed.

Material Removal

The material removal function is critically important to all drilling systems, and many varied and ingenious techniques have been developed. However, all material removal systems can be grouped into a few categories according to the process used. The following categories are suggested:

1. Direct lifting of cuttings or cores

2. Lifting of cuttings by fluid suspension (air, liquids, or foams)

3. Lateral displacement of material (especially in compressible soils)

4. Dissolving of cuttings.

Direct lifting can be accomplished by continuous screw transport using helical flights, by intermittent lifting of buckets, grabs or screws, and by intermittent lifting of core barrels. Continuous flight augers transport cuttings directly from the bit to the surface by screw action. Ideally, flow rate through the screw is equal to production rate at the bit, but in many ground conditions cuttings spill between the outside of the flight and the hole wall, so that the flight tends to recycle cuttings (this is one reason why bristle seals and flight casings have been developed). Continuous flighted systems find their main application in shallow drilling, usually not more than 100-ft (30-m) depth. Intermittent lifting of cuttings after finite intervals of bit penetration can be accomplished with a variety of devices. Bucket augers load directly from the bit, as do the short sections of low-pitch auger flight that accumulate cuttings until lifted clear. Flighted core barrels also load directly, both with core and with cuttings from the annulus between the core and the hole wall. There are also grabs, typically used with cable tool systems, that are lowered into the hole to extract cuttings after the bit has been removed.

Suspension transport is the most widely used and the most broadly applicable method for cutting removal at the present time. In a typical arrangement, fluid is fed continuously down the center of the drill rod or pipe, out past the bit, and back up the annulus between the drill stem and the hole wall. The fluid may be air, water (often with additives to increase density and viscosity), or other liquids (e.g. kerosene or diesel fuel for low temperature operations). The flow velocity (which is controlled by an air compressor or a fluid pump) must be sufficient to suspend and transport the cuttings. This type of system can be applied to almost every type of drilling system, from small hand-held percussive drills to deep oil-well rotary rigs. In a variant of the circulation pattern just described, fluid enters the hole down the annulus and returns up the drill stem, impelled by suction from the return end or by direct pumping into the annulus. When air or untreated water is used as the transport fluid, the discharged fluid with its load of waste products is often discarded, but when treated water or other expensive fluids are used, the discharge is passed through a settling system to remove cuttings and the fluid is then recirculated.

Lateral displacement of surplus material can be applied when rods or tubes are thrust into material that can be moved to accommodate the penetration, either by compaction, by plastic flow, or by absorption of liquefied waste products. Drive sampling, vibratory drilling, and pile driving in soils are examples of processes that require material to displace laterally. When a thermal drill or probe penetrates dense snow on glaciers and polar ice caps, the surplus meltwater can be absorbed and refrozen in the adjacent snow. A similar principle has been suggested for disposal of melted rock produced by thermal drills and tunnel borers, and it appears to be applicable in some rock types.

Solution. The change of solids to a liquid state can provide an attractive alternative to aid in transport or penetration of some materials, e.g. ice and salts. This is particularly true if the minerals require relatively small energy levels for a change of state. This could permit materials to be transported up the hole without the use of more cumbersome mechanical methods such as flight augers, and also eliminate the requirement for pump circulation systems to be designed to handle solid particles.

Hole Wall Stabilization

It is essential to maintain stability of the hole wall while a drilling operation is in progress, and in many cases it is desirable to maintain stability after the completion of drilling. The primary objectives are to prevent wall failure and erosion of the wall by drilling fluids, and also to restrict lateral fluid movement into or out of the hole.

There are three general approaches to stabilization: (a) direct mechanical constraint with a rigid casing, (b) direct constraint with fluids, and (c) treatment of the hole wall material to improve its mechanical properties.

Direct constraint by mechanical means is usually provided by metal pipe placed in close contact with the hole wall. This type of casing can be placed either by driving it with pneumatic casing hammers or large drop hammers, or by drilling it in with a bit set on the bottom of the casing.

Casing can be placed after a hole is completed, or concurrently with a drilling operation. The approach used depends on the drilling equipment used, material properties, and objective of the drilling program. When casing is placed after hole completion, the conditions can vary from stable ground, which causes limited problems, to unstable ground where it is necessary to use high density fluids to maintain an open hole until the casing is placed.

Concurrent placing of casing with the drilling operation involves the progressive or simultaneous advancement of the casing and drill string. The choice of advancing the casing ahead of or behind the drill or sampling tool is controlled by ground conditions and the program objective.

Direct constraint by liquids is employed in many drilling situations when hole wall stability is a problem. A high-density liquid or drilling mud is usually used as a drilling fluid, which loads the hole wall and prevents wall failure. In ice, this technique has been used in deep holes to retard closure of the hole by creep.

Treatment of the ground material usually involves the use of specialized muds, cementing techniques, or freezing. Specialized muds are often used to seal permeable rock types. In some cementing operations, cement grout is forced under pressure into the unstable or permeable soil or rock. The distance to which the ground can be grouted is determined largely by material properties. Freezing operations might be subdivided into active applications, in which previously unfrozen ground is frozen, and passive applications, in which frozen ground is maintained in the frozen state. In all passive applications, and in some active applications, thermal control is achieved most readily by circulating cold drilling fluid in a suspension transport system. In very cold weather, heat exchange between the drilling fluid and ambient surface air can be utilized, but in other circumstances it is necessary to refrigerate the drilling fluid. In some shaft-sinking applications that involve active freezing, freezing pipes may be driven in a ring around the shaft area to freeze the ground ahead of sinking operations. In order to maintain hole wall stability in frozen ground after drilling is completed, it may be necessary to insulate or refrigerate on a long term basis, perhaps using special casing.

With the new rock-melting drills, hole wall treatment is achieved by the melted rock material being displaced laterally into joints and pores of the adjacent material. Upon solidification a very dense and impermeable hole wall liner is formed.

BASIC ENERGY AND POWER REQUIREMENTS

In all drilling operations energy has to be supplied in order to penetrate the ground material and in order to remove surplus material. Energy is also required to lift and lower the drilling equipment in the hole. The rate at which energy has to be supplied determines the power requirements of the drilling system. In many practical drilling systems the inefficiencies and losses represent a significant addition to basic power requirements, but nevertheless it is important to analyze the basic requirements in order to determine how energy and power are distributed among the various elements of the drilling system.

Minimum Energy and Power Requirements for Cutting and Chipping

In a mechanical drilling process a certain amount of energy is needed solely for cutting and chipping the material that is being penetrated. It is convenient to define this energy as the specific energy for cutting, i.e. the work done per unit volume of material cut. The absolute irreducible minimum value for this specific energy is given by the fracture surface energy of the material multiplied by the specific area (area per unit volume) of the cuttings (surface energy represents the energy change when material is cleaved so that some atoms or molecules change from the fully bounded condition of the bulk material to the partially bounded condition of surface material). It is fairly obvious that this minimum specific energy will vary with the size of cuttings, since specific surface area decreases as chip size increases. Taking surface energy as constant for a given material, and specific surface as inversely proportional to a linear dimension of the chip, minimum specific energy is therefore also inversely proportional to chip size, i.e. it is very large when the chips are fine but drops to very low values when the chips are very large.

When it comes to matters of practical determination, surface energy is a somewhat nebulous quantity and it is usual to simply define specific energy for a given cutting or breaking process, e.g. specific energy for indentation, shear cutting, etc. Values are obtained for a given material by measuring the actual work performed by the cutting tool and dividing it by the resulting volume of material removed. For a given material and a given cutting process, specific energy varies with the size of cuttings, as already discussed, with the condition of the material (e.g. temperature, water content, porosity, etc.), with the geometry of the tool (shape, spacing and sequence of cutters), and with the rate of loading or straining (especially if there is a transition from ductile to brittle material response).

If a realistic estimate of specific energy can be made for a cutting process that is to be utilized by a drill, then minimum power requirements for operation of the bit can be calculated. If E_s is the specific energy for cutting, D is hole diameter and R is penetration (feed) rate, then the power required for actually cutting the material, P_c, is:

$$P_c = \frac{\pi}{4} D^2 R E_s \qquad \qquad \text{Eq. (1)}$$

If D is in inches, R is in inches per minute, and E_s is in in.-lbf/in.^{3} (or lbf/in.^{2}), then the required power is:

$$P_c = 1.98 \times 10^{-6} D^2 R E_s \qquad \text{hp} \qquad \qquad \text{Eq. (2a)}$$

If D is in meters, R is in mm/sec, and E_s is in J/m^3 (or N/m^2), then the required power is:

$$P_c = 7.85 \times 10^{-5} D^2 R E_s \qquad \text{kW} \qquad \qquad \text{Eq. (2b)}$$

Frozen soil. There are two main sources for experimental values of E_s for frozen soils: Zelenin (1959, 1968) and Bailey (1967). Zelenin made a major study of the strength and cutting resistance of frozen soils and for his cutting tests he used a large shearing or grooving apparatus and also a drop-wedge for chipping the edge of block samples. His shearing tests were made with drag bits 0.4 to 7.9 in. (10 to 200 mm) wide, cutting at depths from 0.4 to 2.8 in. (10 to 70 mm) at a speed of approximately 1 in./sec (25 mm/sec). For sandy loam at temperatures in the range -1° to -3°C, and at water contents of 18 to 34 per cent, he obtained* values of E_s mainly in the range 300 to 1800 lbf/in.^{2} (2 to 12 MN/m^2). E_s decreased with increasing width of cut, but did not change much with cut depth in the range studied. E_s was a maximum at a certain water content, which probably corresponded to the ice saturation value, and it increased significantly with decreasing temperature (by a factor of 4 as temperature dropped from -1° to -20°C). The drop-wedge, which turned out to have an optimum edge angle close to 30°, gave some extremely low values for E_s, down to about 50 lbf/in.^{2} (0.3 MN/m^2), but these probably resulted from unrealistically favorable situations, since other results ranged up to 1000 lbf/in.^{2} (7 MN/m^2). It was also found that with optimum interaction of multiple cutters, E_s could be lowered to 65 to 85 per cent of the single cutter value. Bailey made shearing experiments by turning cylinders of frozen soil in a lathe, using a variety of small cutting tools that took cuts from 0.02 to 0.2 in. (0.5 to 5 mm) deep. He tested sand, silt, and mixtures of sand and silt, mainly at -3° to -10°C, obtaining values of E_s in the range 400 to 2400 lbf/in.^{2} (2.8 to 16 MN/m^2). E_s decreased with increasing cut depth by 50 to 100 per cent over the size range studied, and also decreased continuously as the tool rake was varied from -20° to +35°. There was a slight increase in E_s as temperature decreased and as dry unit weight increased. Bailey also made experiments in which wedges were indented normally into surfaces of frozen sand and frozen silt at various speeds and temperatures, and with varying wedge angle. Values of E_s varied from about 600 to 7000 lbf/in.^{2} (4 to 48 MN/m^2), but for sand they were typically in the range 600 to 2000 lbf/in.^{2} (4 to 14 MN/m^2) and for silt typically in the range 1000 to 2000 lbf/in.^{2} (7 to 14 MN/m^2). E_s increased as wedge angle increased from 30° to 90°, and tended to decrease when indentation craters were spaced closely enough for interference. For sand, there was not much evidence of significant influence by either temperature or striking velocity, but for silt E_s decreased as striking velocity increased from 4 to 75 ft/sec (1.2 to 23 m/sec) and as temperature

---

*Values of E_s were calculated by us from Zelenin's reported values for cutting force.

decreased down to -30°C, as might be expected for material that exhibits some ductility.

To make order of magnitude calculations from Eq. (2), a value E_s = 1000 lbf/in.^{2} (6.9 MN/m^2) can probably be accepted for drag bit tools working on common frozen soils. A similar value might be taken for indentation cutting if the indentation tool works fast enough to induce brittle fracture, but if there is no brittle fracture (e.g., slow roller bit working on fine-grained soil at high temperature) the calculation, like the drilling operation, is futile. Taking E_s = 1000 lbf/in.^{2} and substituting in Eq. (2), $P_c \approx 0.002\, D^2 R$ hp. If D = 6 in. and R = 100 in./min, $P_c \approx$ 7.2 hp, or if D = 10 in. and R = 60 in./min, $P_c \approx$ 12 hp.

Ice. Shear cutting experiments were made on ice by Zelenin (1959), Bailey (1967) and Peng (1958). Zelenin took cuts 2 in. (50 mm) deep in ice at -1°C, and the specific energy ranged from about 280 lbf/in.^{2} (1.9 MN/m^2) for a cut 2 in. (50 mm) wide to about 700 lbf/in.^{2} (4.8 MN/m^2) for a cut 0.4 in. (10 mm) wide. Bailey took shallow cuts with a lathe at temperatures from -3° to -25°C, finding specific energy values in the range 70 to 700 lbf/in.^{2} (0.48 to 4.8 MN/m^2). Specific energy dropped by a factor of about 5 as cutting depth increased from 0.02 to 0.2 in. (0.5 to 5 mm), but it did not vary much with either temperature or cutting speed (in the range 1 to 10 ft/sec, or 0.3 to 3 m/sec). Variation of tool rake from -20° to +35°C did not seem to have much effect on specific energy. Peng's work appeared rather confused, but from his results Bailey estimated that specific energy was about 200 lbf/in.^{2} (1.4 MN/m^2) at -2°C with cutting depth 0.125 to 0.25 in. (3.2 to 6.4 mm), tool width about 0.5 in. (13 mm), and cutting speed 1 to 4 ft/sec (0.3 to 1.2 m/sec). Bailey (1967) also made wedge indentation experiments on ice, finding specific energy values in the range 70 to 500 lbf/in.^{2} (0.48 to 3.4 MN/m^2) for temperatures in the range -3 to -30°C. There was no convincing evidence of much dependence on either temperature or entry velocity (in the range 3 to 40 ft/sec, or 0.9 to 12 m/sec), but specific energy increased as wedge angle increased from 30° to 90°. Lowest energy values were obtained with blows spaced closely enough for optimum interference.

In Fig. 3 the basic power requirements for cutting or chipping are shown for a range of values of hole diameter, penetration rate, and specific energy. One rather striking feature of this plot is the very modest power requirement for boring small diameter holes at good rates in almost any kind of fine-grained frozen soil or ice. It might be noted that these power estimates assume that the full hole diameter is being cut—for coring, the required power should be lower by a factor of $[1-(D_o/D_i)^2]$, where D_o and D_i are outer and inner diameters of the coring head, respectively.

In laboratory tests on hard rocks, specific energy for indentation tools has been measured by Miller and Sikarskie (1968), Lundquist (1968), and Mellor and Hawkes (1972). Specific energy for indentation with disc cutters has been measured by Bruce and Morrell (1969) and by Rad (1970). The overall range of specific energy values covers more than an order of magnitude, and there is a linear correlation with the uniaxial compressive strength of the material tested (see Mellor, 1972a). The ratio of specific energy to uniaxial compressive strength is mainly between 1.0 and 0.4 (Fig. 4). Basic power requirements for chipping rock with percussive bits or roller bits can be estimated by first estimating the probable limits of specific energy (between 100 and 40 per cent of the uniaxial compressive strength), and then reading power from the appropriate scales of Fig. 3, using multiplying factors of 10 or 100 if necessary (if specific energy for a certain rock is 20,000 lbf/in.^{2}, power can be read from the 200 lbf/in.^{2} scale and multiplied by 100).

Laboratory data on specific energy for drag-bit cutting in hard rock are scarce, but Barker

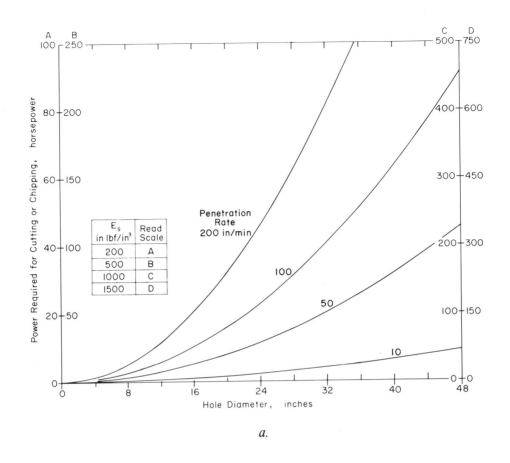

a.

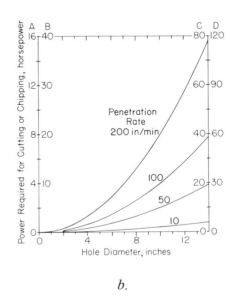

b.

Figure 3. Basic power requirement for cutting or chipping shown for a range of hole diameters, penetration rates, and specific energy levels.

86

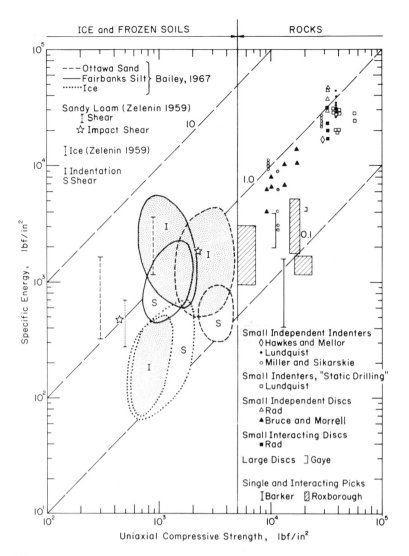

Figure 4. Specific energy consumption from cutting tests on rock, ice and frozen soil plotted against uniaxial compressive strength of the material.

(1964) obtained extremely low values in experiments with large drag bits—specific energies down to 3 per cent of the uniaxial compressive strength of the rock with optimum depth and spacing of cuts.

All available data for specific energy consumption in laboratory cutting tests have been compiled in Fig. 4. Specific energy for cutting of rock, ice and frozen soils is plotted against uniaxial compressive strength on logarithmic scales and a linear correlation is suggested in accordance with findings in the field of rock mechanics. Most of the data lie in a band bounded by $0.1\,\sigma_c < E_s < 1.0\,\sigma_c$, where σ_c is uniaxial compressive strength.

Minimum Energy and Power Requirements for Penetration by Melting

When a bit or probe penetrates a material by melting it completely, the material has to be heated to the melting point and latent heat of fusion for the melted fraction has to be supplied (an alternative for some rocks is to thrust the bit through softened, but not completely melted, material). In addition to the demand for sensible and latent heat, there is unavoidable but unproductive heat flow to the liquid fraction. This last item can become very serious if the drill is immersed in meltwater. Heat losses at the drill head are not easy to estimate in simple terms, especially for ice; a relatively simple analytical scheme for typical rocks has been developed by Murphy and Gido (1973), and a more complete but rather complicated analysis for ice has been made by Shreve (1962). However, for present purposes, which relate to general planning, a first estimate of the lower limit of power requirements can be obtained by assuming efficient heat transfer at the drill tip and ignoring unproductive heat losses to the surrounding material and to the melt.

For melting calculations on frozen materials, it will be assumed that all of the ice in the material to be removed is melted. Thus the minimum thermal power required for melting, P_M, can be expressed as

$$P_M = \frac{\pi}{4} D^2 R \left[m_i \left(S_i \, \Delta\theta + L_i \right) + m_s \, S_s \, \Delta\theta \right] \qquad \text{Eq. (3)}$$

where m_i is mass of ice per unit volume of ground material, m_s is mass of mineral matter (soil grains) per unit volume of ground material, S_i and S_s are specific heats of ice and mineral matter, respectively, L_i is latent heat of fusion for ice, and $\Delta\theta$ is the difference between initial ground temperature and the melting temperature. If the volume fraction of ice is denoted by v_i, then

$$m_i = \rho_i \, v_i \quad \text{and} \quad m_s = \rho_s \, (1 - v_i)$$

where ρ_i is density of ice (0.917 g/cm^3) and ρ_s is density of soil grains ($\approx 2.7 \text{ g/cm}^3$ for common soils).

Since sensible heat is likely to be small relative to latent heat for materials that have high ice content, a fixed value of $\Delta\theta$ can be taken for most calculations that deal with natural frozen ground or natural ice masses. For present purposes $\Delta\theta$ is taken as 5°C. Apparent specific heat of ice at -5°C can be taken as 0.5 cal/g-°C, and latent heat of fusion for phase change at 0°C can be taken as 79.7 cal/g. Specific heat for soil grains can be taken as 0.2 cal/g.

For solid ice, $v_i = 1.0$, and hence

$$P_M = 0.0908 \, D^2 R \quad \text{hp}$$

where D is in inches and R is in in./min. For ice-bearing soils, using the same units,

$$P_M = 1.204 \times 10^{-3} \, D^2 R \, (72.8 \, v_i + 2.7) \quad \text{hp}$$

In Fig. 5 the minimum power requirements for melting are plotted as a function of diameter for various penetration rates and ice contents. If this graph is compared with Fig. 3, it can be seen straight away that thermal drilling makes very much heavier power demands than direct mechanical drilling for the penetration process.

Eq. (3) implies that penetration rate is directly proportional to power density, i.e. power divided by the working area of the boring head. However, there are practical limits to the power density that can be achieved with an electrical heater that has to have a reasonable working life

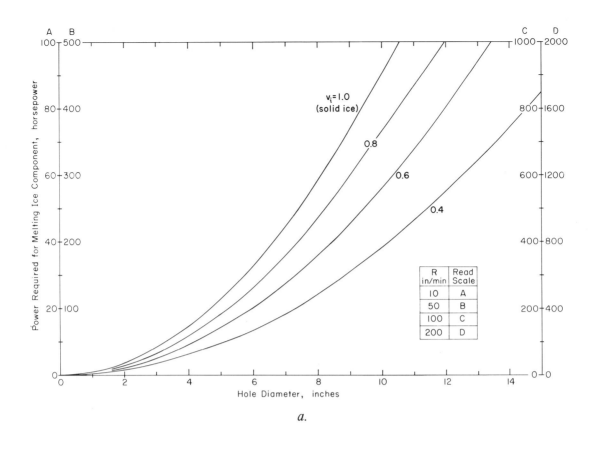

a.

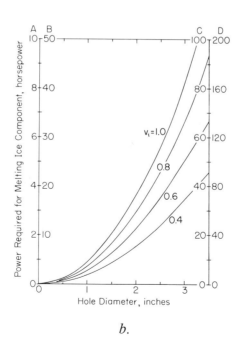

b.

Figure 5. Minimum power requirements for melt penetration of ice and frozen soils, shown as a function of hole diameter, penetration rate, and ice content.

89

(and also limits to the power density that can be usefully employed). Shreve and Sharp (1970) addressed this problem, and developed a hot-point that had a working life better than 1000 hr at a power density of 1.2 MW/m^2. Similar efforts have been made in France, and power densities up to 3.25 MW/m^2 have been employed effectively (F. Gillet, personal communication). The "Subterrenes" under development at the Los Alamos Scientific Laboratory operate at high temperatures, but their power densities are in the same range as those of ice drills–existing models have worked in the range 0.3 to 2.5 MW/m^2 (Armstrong, 1974), and requirements up to 5 MW/m^2 have been noted.

With an effective limit on power density, there is a limit to the attainable penetration speed with a thermal drill. Eq. (3) can be rewritten for the limiting case in terms of maximum penetration rate R_{max} and maximum power density $(P/A)_{max}$:

$$R_{max} = \frac{(P/A)_{max}}{[m_i\,(S_i\,\Delta\theta + L_i) + m_s\,S_s\,\Delta\theta]} \qquad \text{Eq. (4)}$$

In the case of solid ice at -5°C, the maximum penetration rate for a useful power density of 3 MW/m^2 is 9.5 mm/sec, or 1.87 ft/min. In other words, thermal drills of the type used so far do not appear to have the potential for development into very rapid ice drills (mechanical ice drills have achieved penetration rates an order of magnitude higher than the present limit for electrothermal drills).

Minimum Power Requirements for Removal of Material from Open Hole

The basic power demands for typical penetration processes (excluding losses and inefficiencies) are not much affected by hole depth but this is not the case for removal of cuttings, core or waste. The *minimum* amount of energy required to remove waste from an open hole of given depth is equal to the weight of material multiplied by the height of lift. If it is assumed that waste material is removed from the hole at the same rate at which it is produced by the penetration process, then the *minimum* power requirement for lifting material, P_L, is

$$P_L = \frac{\pi}{4}D^2\,R\gamma_g\,h \qquad \text{Eq. (5)}$$

where γ_g is unit weight of the ground material in place, and h is the hole depth. With D in inches, R in in./min, γ_g in lbf/ft^3 and h in ft,

$$P_L = 1.376 \times 10^{-8}\,D^2 R\,\gamma_g\,h \qquad \text{hp}$$

This relationship is shown graphically in Fig. 6, and it can be seen that the basic power requirement for lifting cuttings is trivial for all but very deep holes and very large diameter holes.

Minimum Power Requirements for Hoisting the Drill String

When the drill string is being removed from the hole, either for core removal or at the end of the operation, it is usually desirable to hoist at an appreciable speed, and this can make a significant power demand. The *minimum* power requirement for hoisting, P_H, is determined by the submerged weight of the suspended string and the hoisting speed, R_H:

$$P_H = whR_H \qquad \text{Eq. (6)}$$

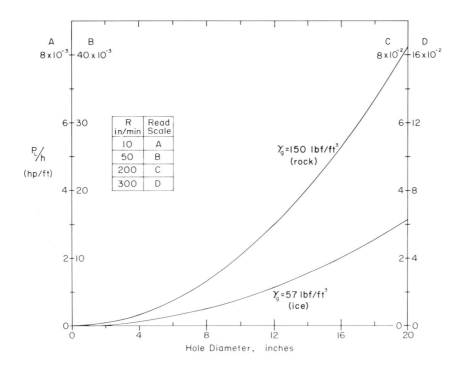

Figure 6. Basic power requirements for continuous lifting of cuttings shown as a function of hole diameter, penetration rate, hole depth, and material type.

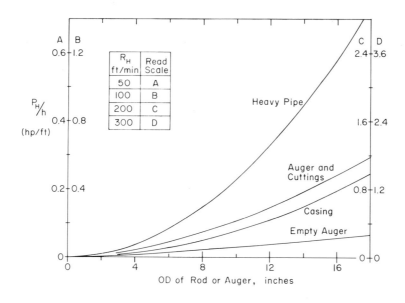

Figure 7. Minimum power requirements for hoisting drill pipes, casing and augers at various rates in a dry hole.

91

where w is the submerged weight per unit length of the drill string and h is the length of the string. For a drill string that is immersed in a viscous fluid, there is an additional power requirement for overcoming fluid resistance, which increases with increasing hoisting rate.

For purposes of illustration, power requirements for hoisting in open hole will be considered. The weight per unit length of drill stem is a function of rod diameter. Weight per unit length would be proportional to diameter squared for geometrically similar rods or augers; this is approximately the case for drill pipe and casing, but continuous flight augers increase in unit weight at a lower rate because the flights become wide relative to the core rod as diameter increases. It will be assumed here that the weight of heavy drill pipe in air is $1.5D^2$ lbf/ft, the weight of casing is $0.5\,D^2$ lbf/ft, and the weight of continuous flight auger is $D^{1.3}$ lbf/ft, where D is in inches and the relations are restricted to the common range of drill sizes. Figure 7 gives power requirements as a function of diameter and hoisting speed for pipe, casing, empty auger, and auger jammed full of cuttings. In many drilling systems this function requires the most power.

Assessing Power Requirements for Complete Drilling Systems

The basic power requirements for a complete drilling system can be analyzed by going through a series of exercises similar to those just outlined. To these minimum estimates must be added the power needed to support the inefficiencies of practical processes and equipment.

Estimation of efficiencies and power losses is an important topic, since mechanical efficiency is often traded for convenience in practical operations. One way to arrive at estimates of power losses is to draw up energy budgets for actual working systems, comparing the overall input of work with the energy expended usefully.

In assessing the partitioning of power input for a drilling system, it has to be recognized that not all functions are performed concurrently, so that a single power source can sometimes be applied to two or more functions in sequence. For example, bit rotation and chip clearance can cease when rod is being hoisted.

MEASURED PENETRATION RATES FOR EXISTING DRILLING TOOLS

The following notes give examples of actual penetration rates for various types of existing equipment. Most of the information is taken from an unpublished report by Mellor *et al.* (1973), which illustrates many of the pieces of equipment that are referred to.

Ice. Drilling in ice presents no great problem if the equipment is properly designed and operated, but some projects have foundered because of inability to drill ice. Well-designed drag bits are the simplest and probably the most efficient tools for cutting ice, as they require very little down-thrust, modest torque, and no percussion. If the ice is perfectly clean and of zero salinity, drag bits do not require carbide tips or hard-facing, although some surface hardening is desirable. A slight amount of rock dust can create wear problems (Abel, 1961), as can inclusions of precipitated salt crystals (G. Lange, unpublished).

Small-diameter holes can be drilled with simple hand equipment at rates that are acceptable for some purposes. The 1.5 in. (38 mm) diameter USA CRREL ice auger (essentially a ship auger with modified tip), rotated by a hand brace, can drill to 3 ft (1 m) at rates from 1.6 to 2.95 ft/min (8.1 to 15 mm/sec) (Kovacs, 1970; Sellmann and Mellor, 1974). With an electric or gasoline power-drive the same tool can penetrate to 3 ft (1 m) at rates from 3.2 to 7.6 ft/min (16 to 39mm/sec) (Kovacs, 1970; Kovacs et al., 1973; Sellmann and Mellor, 1974). Like any auger, this tool can be overdriven so that cuttings jam in the flight, and care must be exercised to match penetration rate with cutting clearance rate.

A simple 1.5 in. (38 mm) diameter flight auger fitted with improved bits has drilled ice at rates up to 4.4 ft/min (22 mm/sec) when driven by a hand brace, and at rates up to 15.1 ft/min (77 mm/sec) when driven by electric hand drills (Sellmann and Mellor, 1974). A 2.2 in. (56 mm) diameter variant penetrated at rates up to 10.4 ft/min (53 mm/sec).

The USA CRREL 3 in. (76 mm) coring auger is sometimes used solely for drilling holes, producing a hole of approximately 4.4 in. (112 mm) diameter. When turned by a hand brace, penetration rates of 0.8 to 1.2 ft/min (4 to 6 mm/sec) have been measured; when the same tool was turned with a T-handle the rates dropped to 0.43 to 0.61 ft/min (2.2 to 3.1 mm/sec) (Kovacs, 1970). With a gasoline drive, rates of 3.0 to 3.5 ft/min (15 to 18 mm/sec) have been measured (Kovacs, 1970). With electric drive, penetration rate has been measured at 2.4 to 4.0 ft/min (12 to 20 mm/sec) (Kovacs, 1970) and 5.4 to 5.6 ft/min (27 to 28 mm/sec) (Kovacs et al., 1973).

A Russian hand-operated cutting ring device, used for coring or hole making, produces an annular hole 8.8 in. (224 mm) O.D. and 7.25 in. (184 mm) I.D. at the rate of 0.2 to 0.33 ft/min (1 to 1.7 mm/sec) (Cherepanov, 1969). Drilling through 7 ft (2 m) thick first-year sea ice takes 30 to 45 min (R. Ramseier, personal communication).

A wide variety of commercial earth augers, or post-hole diggers, have been adapted for drilling ice, especially for the use of ice-fishermen. They commonly have diameters ranging from about 4 in. to 9 in. (0.1 to 0.23 m), and are normally intended for drilling to depths of only a few feet, although the writers have drilled to 16 ft (5 m) with 9 in. (0.23 m) diameter hand-held gasoline-powered augers. Kovacs (1970) has driven an 8 in. (0.2 m) diameter earth auger with various gasoline and electric drive units at a penetration rate of 1.2 ft/min (6.1 mm/sec). The writers have drilled numerous 9 in. (0.23 m) diameter holes at somewhat higher rates (approximately 3 ft/min) with freshly sharpened ice augers, and ice-fishermen have claimed rates approaching 5 ft/min (25 mm/sec) with 9 in. (0.23 m) diameter augers, and 6 ft/min (30 mm/sec) with 7 in. (0.18 m) diameter augers. In controlled tests, a 9 in. (0.23 m) diameter auger penetrated at 5.3 to 7.5 ft/min (27 to 38 mm/sec), and a 5.5 in. (0.14 m) diameter auger penetrated at 5.4 to 7.5 ft/min (27 to 38 mm/sec) (Kovacs et al., 1973).

Shothole drills developed for underground mining have been used to drill ice with a minimum of modification. Rausch (1958) drilled 1.75 in. (44 mm) diameter shotholes in ice with pneumatic rotary-percussive mining drills, achieving penetration rates of 5 ft/min (25 mm/sec). Abel (1961) used percussive augers to drill 1.75 in. (44 mm) diameter shotholes, obtaining overall penetration rates better than 5 ft/min (25 mm/sec) for 8 ft (2.4 m) long holes. He also used a hand-held electric powered auger to drill 2 in. (51 mm) diameter holes at 5 ft/min (25 mm/sec). McAnerney (1968) used a hydraulically driven hand-held coal auger for boring 1.75 in. (44 mm) diameter shotholes in frozen silt and ice, obtaining penetration rates up to 11.75 ft/min (60 mm/sec) in lenses of pure ice. Kovacs et al. (1973) drove 1.75 in. (44 mm) diameter face augers and roof-bolt augers with electric drills, and achieved penetration rates up to 9.5 ft/min (48 mm/sec).

The writers have drilled with hand-held electrically driven 3 in. (76 mm) diameter augers to depths of 55 ft (17 m) using a variety of bits. With good bits, short-term penetration rates (4-ft increments) of 15 ft/min (76 mm/sec) were attainable. Controlled tests with similar tools gave penetration rates up to 14 ft/min (71 mm/sec) (Kovacs *et al.,* 1973). Kovacs (1974) has developed a lightweight 3 in. (76 mm) diameter auger that penetrates at up to 10.4 ft/min (53 mm/ sec) with an electric drive unit. Similar rates of 3.4 to 13.9 ft/min (17 to 71 mm/sec) were reported for small-diameter auger drills in river and sea ice by Russian workers (Nikolaev and Trubina, 1969).

From the foregoing performance records it is clear that hand-held drive units are perfectly adequate for supplying the power, torque and thrust required for drilling holes up to 9 in. (0.23 m) diameter at fully acceptable rates in ice. However, frame-mounted units are required for hoisting and lowering when holes have to be drilled to considerable depths. The higher power that is usually available in a frame-mounted unit does not permit any significant increase in penetration rate over hand-held units, since cutting clearance sets a limit (an inept operator can twist off the auger stem if a highly powered unit is over-driven so that cuttings are jammed).

The U.S. Navy used a trailer-mounted drilling unit (approximately 5 tons) for experimental drilling in sea ice. Maximum penetration rate was 8 ft/min (41 mm/sec) with a 4.75 in. (0.12 m) diameter tricone roller bit, and 1 ft/min (5 mm/sec) with a 14 in. (0.36 m) O.D. (12 in. or 0.3 m I.D.) coring bit (Hoffman and Moser, 1967). Tests were also made with a 10 in. (0.25 m) diameter auger, which penetrated at 6 ft/min (30 mm/sec) (Beard and Hoffman, 1967).

For deep drilling in Greenland and Antarctica, USA CRREL has used an electromechanical coring drill. The drill bit had a maximum outside diameter of 6.13 in. (156 mm) and minimum inside diameter of 4.50 in. (114 mm). Penetration rates have been in the range of 0.12 to 0.66 ft/min (0.61 to 3.4 mm/sec) (Ueda and Garfield, 1968, 1969a, 1970).

A lightweight (500 lb or 230 kg) powered ice-coring auger developed by the former Arctic Construction and Frost Effects Laboratory (ACFEL) penetrated at 0.67 to 1.0 ft/min (3.4 to 5.1 mm/sec), taking 3 in. (76 mm) diameter core and making a 4.75 in. (121 mm) diameter hole (ACFEL, 1954).

Thermal drills have also been used for boring holes in ice, although they are very inefficient in energetic terms compared with mechanical drills. Electrical hot-point drills usually penetrate at rates not exceeding 60 to 80 per cent of the theoretical rates calculated on the basis of melting with no heat loss. Theoretical penetration rates for lossless melting were given earlier, and some practical heat losses are discussed by Aamot (1967a, 1968). To give an idea of penetration rate, a 2 kW (2.7 hp) electric hot-point can readily bore 2 in. (51 mm) diameter hole at 0.33 ft/min (1.7 mm/sec). Shreve and Sharp (1970) achieved rates up to 0.49 ft/min (2.5 mm/sec) with 2.1 kW on a 2 in. (51 mm) diameter hot-point, while Stacey (1960) reached 0.63 ft/min (3.2 mm/ sec) at 2.3 kW (3.1 hp) and 0.5 ft/min (2.5 mm/sec) at 1.8 kW (2.4 hp) for the same size bit. LaChapelle (1963) drilled at 0.30 to 0.33 ft/min (1.5 to 1.7 mm/sec) with 0.22 kW (0.3 hp) on a 0.71 in. (18 mm) diameter hot-point. The 3.625 in. (92 mm) diameter Philberth probe penetrated at 0.16 ft/min (0.81 mm/sec) with 3.68 kW (4.9 hp) input in Greenland (Aamot, 1967b).* One of us has bored 0.73 in. (19 mm) diameter holes to depths of 200 ft (61 m) at a rate of 0.27 ft/min (1.4 mm/sec) with a 0.25 kW (0.34 hp) electric hot-point. W. Tobiasson

---

*Philberth (this symposium) gives 0.11 ft/min (0.56 mm/sec) as the maximum rate of the 3.7 kW probe.

(personal communication) has bored with a 0.5 kW (0.67 hp), 1.25 in. (32 mm) diameter hot-point at rates of 0.15 and 0.22 ft/min (0.76 to 1.1 mm/sec). On a larger scale, the 6.4 in. (0.16 m) diameter USA CRREL thermal coring drill has penetrated at rates from 0.126 ft/min (0.64 mm/sec) in $0^{\circ}C$ ice to 0.104 ft/min (0.53 mm/sec) in ice at $-28^{\circ}C$, the input power ranging from 3.5 to 4.0 kW (4.7 to 5.4 hp) (Ueda and Garfield, 1969b). Russian electrothermal penetrators have drilled at 0.38 to 0.49 ft/min (1.9 to 2.5 mm/sec) with 1 to 2 kW (1.3 to 2.7 hp) on a tip diameter of 1.6 in. (40 mm) and at 0.38 to 0.55 ft/min (1.9 to 2.8 mm/sec) with 3 to 4 kW on a tip diameter of 3.1 in. (80 mm) (Korotkevich and Kudryashov, this symposium). Russian electro-thermal corers have drilled at 0.16 to 0.25 ft/min (0.83 to 1.25 mm/sec) with 1.5 to 2.2 kW (2 to 3 hp) on a wedge-profile annulus of 3.5 in. (88 mm) inside diameter and 4.4 in. (112 mm) outside diameter, and also at 0.08 to 0.11 ft/min (0.42 to 0.56 mm/sec) with 3.5 kW (4.7 hp) on a flat-base annulus of 5.1 in. (130 mm) inside diameter and 7 in. (178 mm) outside diameter (Korotkevich and Kudryashov, this symposium). The French "bare-wire" thermal corer is reported to have achieved rates up to 0.33 ft/min (1.7 mm/sec) with about 4.1 kW (5.4 hp) on a head boring 5.5 in. (0.14 m) diameter hole and taking 4 in. (0.1 m) diameter core (F. Gillet *et al.,* this symposium).

Lightweight steam drills have been developed for boring in ice; a recent design (Hodge, 1971) has bored 1 in. (25 mm) diameter hole to 26 ft (7.9 m) depth at 1.8 ft/min (9.1 mm/sec), and 2 in. (51 mm) diameter hole at 0.49 ft/min (2.5 mm/sec). In an earlier effort, Howorka (1965) drilled 0.8 in. (21 mm) diameter hole to 26 ft (8 m) with a 0.1 in. (2.5 mm) diameter steam nozzle at a rate of 0.87 ft/min (4.4 mm/sec).

Browning and Ordway (1963) used a flame jet to drill 7.5 in. (0.19 m) diameter hole in ice at 2.9 ft/min (15 mm/sec).

Frozen fine-grained soils. Drilling in frozen soil is often considered to be a difficult task equivalent to hard-rock drilling, but in fact holes up to 4.5 in. (0.11 m) diameter or more can be drilled in frozen fine-grained soils with hand-held units.

The writers have drilled 3 in. (76 mm) diameter holes in frozen silts with continuous-flight, gasoline-powered augers at rates up to 7 ft/min (36 mm/sec), with penetration rates of 6.5 ft/min (33 mm/sec) readily attainable. They have also drilled 4.4 in. (0.11 m) diameter hole with the USA CRREL 3 in. (76 mm) coring auger at short-term penetration rates of approximately 12 ft/min, or 61 mm/sec (appreciably faster than the same tool drilling in solid ice). McAnerney (1968) drilled 1.75 in. (44 mm) diameter holes in frozen silt with a hydraulic, hand-held auger at rates ranging from 2.2 to 11.75 ft/min (11 to 60 mm/sec); the lowest rates were in soil at temperatures close to the melting point, and the highest rates were in cold soil ($17^{\circ}F$, or $-8.3^{\circ}C$) and in ice lenses.

In recent development work, 1.5 in. (38 mm) diameter augers have been driven with a hand brace in frozen silt, achieving penetration rates up to 2.4 ft/min (12 mm/sec) (Sellmann and Mellor, 1974). With electric drill drive units, the same hand augers penetrated frozen silt at rates up to 7.5 ft/min (38 mm/sec).

Heavy powered augers and rotary drilling systems are widely used for shothole drilling and for setting posts and piles. Lange (1964) gives some short-term penetration rates for mine shot-hole drills working in frozen sand. A 50 hp (37 kW) auger drilled 6 in. (0.15 m) diameter shot-holes up to 100 ft (30 m) long at 6.7 ft/min (34 mm/sec), while a 100 hp (74.6 kW) auger drilled 9 in. (0.23 m) diameter holes up to 90 ft (27 m) deep at 4 ft/min (20 mm/sec). A 215 hp (160 kW) rotary rig with air circulation (Chicago Pneumatic 650) drilled 8.25 in. (0.21 m) diameter holes at 6 to 7 ft/min (30 to 36 mm/sec) with drag bits. A Failing 43 rotary drill with air circulation drilled 6 in. (0.15 m) diameter holes in frozen silt with bladed drag bits at 7 to 12 ft/min (36 to 61 mm/sec), with 9.25 ft/min (47 mm/sec) the most frequent rate (Mellor, 1971). Large-diameter augers, such as the Williams auger, do not normally have continuous flight, and cutting removal is cyclic. This results in low penetration rates overall; McCoy (1960) gives 14 to 16 ft/hr (4.3 to 4.9 m/hr) for 12 in. (0.3 m) diameter holes and 12 ft/hr (3.7 m/hr) for 24 in. (0.61 m) diameter holes in frozen peat, gravel and silt. Roller rock bits have sometimes been used for drilling frozen silts, but they are usually very ineffective.

Percussive rock drills are occasionally used for frozen fine-grained soils. McAnerney (1968) used a rotary-percussive air-leg rock drill with liquid circulation to bore 1.75 in. (44 mm) diameter shotholes in frozen silt, and achieved penetration rates of 0.7 ft/min (3.6 mm/sec). A rotary-percussive rock drill with 3 in. (76 mm) diameter bit and air circulation (Gardner Denver 123J) was used for shothole drilling in frozen ground during blasting trials by DuPont (TAPS, 1969). Average penetration rate for a mixed silt/gravel section was 4.5 ft/min (23 mm/sec), with maximum rate of 9 ft/min (46 mm/sec), and it was noted that drilling appeared to be faster in the gravel than in the silt.

Open-end pipe of 6 in. (0.15 m) outside diameter has been driven into frozen silt and sand at rates of 30 ft/min (152 mm/sec) with a high-frequency vibratory unit (Huck, 1969). A low-frequency percussive tool (Ingersoll-Rand Hobgoblin) has been used to drive 4 in. (0.1 m) diameter solid steel rod into frozen silt at 2.3 ft/min (12 mm/sec) with a chisel point and 2.8 ft/min (14 mm/sec) with a moil point (Mellor, 1972b).

McAnerney (1968) used a steam point to drill small-diameter shotholes in frozen silt, achieving penetration rates as high as 4.5 ft/min (23 mm/sec), with an average rate of 3.3 ft/min (17 mm/sec). Browning and Ordway (1963) used a flame jet to drill frozen silt, obtaining penetration rates of 1.1 ft/min (5.6 mm/sec) for 6 in. (0.15 m) diameter hole, 0.67 ft/min (3.4 mm/sec) for 7 in. (0.18 m) diameter hole, and 0.375 ft/min (1.9 mm/sec) for 8 in. (0.2 m) diameter hole. Browning and Fitzgerald (1964) used a redesigned flame jet in frozen silt, and reached penetration rates of 1 ft/min (5.1 mm/sec) for 8 and 9 in. (0.2 and 0.23 m) diameter hole, and up to 1.1 ft/min (5.6 mm/sec) for 7 in. (0.18 m) diameter hole.

It is understood that in laboratory tests at the Los Alamos Scientific Laboratory, very cold frozen silt (-73° and -143°C) was penetrated by a 3 in. (75 mm) diameter high-temperature electrical hot-point at rates up to 0.028 ft/min (0.14 mm/sec) with a power of 6.7 kW (9 hp) and a thrust of 1000 lbf (4.5 kN).

Frozen tills and gravels. When frozen ground contains pebbles and cobbles that are large relative to the cutting tools and the hole diameter, the nature of the drilling problem changes, since these pieces of hard rock have to be cut to permit penetration and removal of cuttings. Thus the drilling of frozen gravels and tills generally calls for rock drilling techniques and equipment.

Rotary drilling systems with roller bits and air circulation (Chicago Pneumatic T-650) have given penetration rates of 2.5 ft/min (13 mm/sec) for 8 in. (0.2 m) diameter hole in frozen gravel (Mellor and Sellmann, 1970). Lange (1968) tested a rotary drilling system (Failing 43) with liquid circulation in a till consisting of frozen clay with cobbles. Several types of drag bits and roller bits were tested for a range of rotational speed and bit loads. Penetration rate increased with increasing rotational speed and increasing bit load, with values ranging up to 2.5 to 3.5 ft/min (13 to 18 mm/sec). Some of the drag bits reached rates of 4 to 6 ft/min (20 to 30 mm/sec), but these rates could not be sustained. A rate of 1.5 ft/min (7.6 mm/sec) was a reasonable limit for efficient removal of cuttings.

Lange (1968) also tested augers in frozen till and obtained penetration rates up to 4.6 ft/min (23 mm/sec) with 6.25 in. (0.16 m) diameter bits. However, the high penetration rates (3 to 4 ft/min, or 15 to 20 mm/sec) resulted in undue tooth breakage and excessive torque on the auger stem, and 1.5 ft/min (7.6 mm/sec) was considered to be the maximum rate for effective cutting clearance. Lange (1973), using a Williams auger (4D-50, capacity: 36 in. hole to 50 ft) in frozen gravel, obtained an average penetration rate of 0.16 ft/min (0.81 mm/sec) in a 16 in. (0.41 m) diameter hole 48 ft (15 m) deep. Similar rates were also obtained with a large rotary Failing 1500, drilling 16 in. (0.41 m) diameter hole.

Abel (1960) used percussive rock drills for tunneling in frozen gravel. The penetration rate of airleg drills with 1.625 in. (41 mm) diameter bits and a frequency of 2000 blows/min (33 Hz) averaged 2.38 ft/min (12 mm/sec). Another drill with the same diameter bit and a frequency of 3000 blows/min (50 Hz) averaged 1.33 ft/min (6.8 mm/sec). Abel also tested 1.485 in. (38 mm) diameter diamond drills, achieving penetration rates that averaged 0.375 ft/min (1.9 mm/sec) for both tapered blast-hole bits and coring bits. Cooled diesel fuel was used as drilling fluid for the diamond drills.

Core barrels with an outside diameter of 4.5 in. (0.11 m) have been driven into frozen gravel at rates of 6 ft/min (30 mm/sec) with a high-frequency vibratory unit (Huck, 1969). A low-frequency percussive unit (Ingersoll-Rand Hobgoblin) has driven 4 in. (0.1 m) diameter solid steel rod into frozen gravel at 0.31 ft/min (1.6 mm/sec) with a chisel point and approximately 0.25 ft/min (1.3 mm/sec) with a moil point (Mellor, 1972b).

Browning and Fitzgerald (1964) drilled frozen gravel with a flame jet, producing 1 ft (0.3 m) diameter hole at a penetration rate approaching 3 ft/min (15 mm/sec).

SPECIFIC ENERGY DATA FOR PENETRATION PROCESSES

Measured Specific Energy for Drag-Bit Penetration

With an operating rotary drill it is awkward to find the process specific energy for cutting, as the total power input covers cutting clearance, hole-wall friction, and mechanical losses as well as the penetration process. However, some reasonably reliable values have been obtained for small drills by measuring power consumption with and without active penetration.

Ice. Kovacs *et al.* (1973) tested a variety of augers and auger bits in ice, obtaining values of overall specific energy for each drill and calculating values of process specific energy for the electrically driven drills. The best values of process specific energy, in the range 100 to 140 lbf/in.^{2} (0.7 to 1.0 MN/m^2), were obtained with two different designs of a 3.25 in. (83 mm) diameter auger bit. Commercial coal bits of 1.75 in. (44 mm) diameter were much less efficient, turning in process specific energy values in the range 400 to 1500 lbf/in.^{2} (2.8 to 10 MN/m^2). The standard USA CRREL 3 in. (76 mm) coring auger had a process specific energy of 350 lbf/in.^{2} (2.4 MN/m^2) based on the volume of ice actually cut, and an effective value of 180 lbf/in.^{2} (1.2 MN/m^2) based on the total hole volume (including core). The standard USA CRREL 1.5 in. (38 mm) diameter ship auger had specific energies in the range 340 to 880 lbf/in.^{2} (2.3 to 6.1 MN/m^2). For overall specific energy, the best values were turned in by two commercial gasoline-powered augers designed for ice-fishermen. A 5.5 in. (0.14 m) diameter auger with a 1 hp (0.75 kW) engine gave an overall value of 185 lbf/in.^{2}), while a 9 in. (0.23 m) diameter auger with a 3 hp (2.2 kW) engine gave typical values from 210 to 300 lbf/in.^{2} (1.4 to 2.1 MN/m^2). Best overall values for electrically driven units were around 300 lbf/in.^{2} (2 MN/m^2).

Sellmann and Mellor (1974) made tests in ice with 1.5 to 2.2 in. (38 to 56 mm) diameter augers, and found best values of process specific energy around 300 lbf/in.^{2} (2.1 MN/m^2), with other values ranging up to 500 lbf/in.^{2} (3.5 MN/m^2) or so. Overall specific energy was in the range 500 to 1200 lbf/in.^{2} (3.4 to 8.3 MN/m^2).

Kovacs (1974) tested a 3 in. (76 mm) diameter ice auger and obtained an extremely low value for process specific energy of 57 lbf/in.^{2} (0.39 MN/m^2) (better than the best values from laboratory experiments), with an overall specific energy of 240 lbf/in.^{2} (1.7 MN/m^2).

From the test results it seems that a process specific energy of 100 lbf/in.^{2} (0.7 MN/m^2) is not an unreasonable design goal, even for small drills that cannot enjoy the scale advantages of larger machines. To put this in perspective, a process specific energy of 100 lbf/in.^{2} (0.7 MN/m^2) for ice represents a dimensionless performance index (see Mellor, 1972a) of about 0.1, i.e. the specific energy is about 10 per cent of the uniaxial compressive strength of the material. For overall specific energy, 200 to 300 lbf/in.^{2} (1.4 to 2.1 MN/m^2) seems a reasonable design goal, with lower values more readily attainable on larger drills. In rock drilling research there is a rule-of-thumb that gives a dimensionless performance index of about 0.3 as the practically attainable lower limit for very efficient drills, and present indications are that this rule is not unreasonable for ice.

Frozen fine-grained soil. Sellmann and Mellor (1974) tested small electrically driven augers in frozen silt and obtained process specific energy values in the range 900 to 1600 lbf/in.^{2} (6.2 to 11.0 MN/m^2), with overall values in the range 1500 to 2300 lbf/in.^{2} (10 to 16 MN/m^2).

In undocumented field tests the authors obtained overall specific energy values down to 3300 lbf/in.^{2} (23 MN/m^2) for 3 in. (76 mm) diameter gasoline-powered augers working in permafrost. We also obtained more favorable values boring 4.4 in. (0.11 m) diameter hole in frozen silt with the USA CRREL 3 in. (76 mm) coring auger powered by a gasoline unit. Basing overall specific energy on the volume of material actually cut, values down to 1700 lbf/in.^{2} (12 MN/m^2) were obtained, while effective overall specific energy based on total hole volume dropped as low as 900 lbf/in.^{2} (6.2 MN/m^2).

In normal operation, large industrial drills tend to work less efficiently[*]. For example, Lange (1964) observed a 50 hp (37 kW) auger drilling 6 in. (0.15 m) diameter hole with overall specific energy consumption of 8700 lbf/in.^{2} (60 MN/m^2), and a 100 hp (75 kW) auger drilling 9 in. (0.23 m) diameter hole with overall specific energy consumption of 13,000 lbf/in.^{2} (90 MN/m^2).

However, other types of very large rotary-cutting devices employing large drag bits have demonstrated much lower values of specific energy under frozen-silt field conditions. For example, large disc saws have cut with overall specific energy as low as 900 lbf/in.^{2} (6.2 MN/m^2) (Mellor, 1975), a tunneling machine has had values down to 700 lbf/in.^{2} (4.8 MN/m^2), a large rotary trencher has given the spectacularly low value of 180 lbf/in.^{2} (1.2 MN/m^2), and a large miller/planer has given values of *process* specific energy down to 720 lbf/in.^{2} (5 MN/m^2) (Mellor, 1972c).

---

*Note added in proof: Our recent studies on the Alaska pipeline revealed that large drilling rigs (rotary, rotary-percussive, percussive) working in frozen soils gave penetration rates and energetic efficiency values far lower than typical values listed in this paper.

There is obviously a lot of scope for design improvements in this material. In some cases attempts to combat abrasion and impact problems have led to poor tool geometry, but there are other factors involving both the kinematics and dynamics of the machines.

Measured Specific Energy for Thermal Penetration

The lower limit of specific energy consumption for thermal penetration of ice and ice-bonded soils is set by the latent heat, ambient temperature, ice content, etc., as already discussed. Putting this limiting value in the same units as are used for mechanical systems, the specific energy consumption for complete melting of solid ice from -5°C is 4.58 x 10^4 lbf/in.^{2} (316 MN/m^2). For frozen soils the corresponding value is approximately proportional to the volumetric ice content for soils that are close to saturation. In operating drilling systems the process specific energy consumption exceeds the theoretical value by an amount that is largely dependent on the power density, the penetration rate, and convective losses, while the overall specific energy consumption is dependent additionally on losses between the energy input source and the melting element. There may also be some question as to whether specific energy should be based on actual hole diameter or the drill diameter.

Electrical drills give the best idea of process specific energy for penetrating ice, since they are not subject to much line loss. Taking some of the penetration rates given in another section of this paper and neglecting bore enlargement, examples can be calculated. A 2 kW (2.7 hp) hot-point boring 2 in. (51 mm) diameter hole at 0.33 ft/min (1.7 mm/sec) gives a specific energy of 8.54 x 10^4 lbf/in.^{2} (589 MN/m^2), or a melting efficiency of 54 per cent. The 3.625 in. (92 mm) diameter Philberth probe penetrating at 0.16 ft/min (0.81 mm/sec) with 3.68 kW (4.9 hp) input gives a specific energy of 9.86 x 10^4 lbf/in.^{2} (680 MN/m^2), or a melting efficiency of 46 per cent. A 0.25 kW (0.34 hp) hot-point of 0.73 in. (19 mm) diameter penetrating at 0.27 ft/min (1.4 mm/sec) gives a specific energy of 9.79 x 10^4 lbf/in.^{2} (675 MN/m^2), or a melting efficiency of 47 per cent.

Shreve and Sharp (1970) obtained a melting efficiency of 75 per cent, LaChapelle (1963) had a melting efficiency of 59 per cent, and Stacey (1960) reached 86 to 88 per cent, all with electrical hot-points.

According to data on Russian electrothermal drills given at this meeting (Korotkevich and Kudryashov), best values of useful specific energy for the small penetrator (1.6 in., or 40 mm, diameter) and the small corer (4.4/3.5 in., or 112/88 mm) working in 0°C ice were about 6 x 10^4 lbf/in.^{2} (400 MN/m^2) and 7 x 10^4 lbf/in.^{2} (500 MN/m^2), respectively. These values represent melting efficiencies of about 74 and 63 per cent, respectively. For the large corer working in ice at temperatures between -28° and -57°C, best values of specific energy were also about 7 x 10^4 lbf/in.^{2} (500 MN/m^2), which represents melting efficiencies in the range 75 to 86 per cent. Results given for the large penetrator (3.1 in., or 80 mm, diameter) working in ice at -19° to -28°C are questionable, as they seem to imply melting efficiencies in excess of 100 per cent. Best reported results for the French thermal corer working in Adélie Land [Coast] (Gillet *et al.,* this symposium) also seem on the optimistic side; 6 m/hr penetration with 4.05 kW (cutting 0.102 m core and 0.14 m hole) in ice at about -14°C implies a melting efficiency of 99 per cent.

The efficiency of a steam drill is more difficult to work out, but Howorka (1965) gave some values for his equipment. About 50 per cent of the input energy was lost between the burner and the boiler output (this has to be compared with the efficiency of an electrical generator). Of the energy put out by the boiler 56 per cent went into line loss, and 44 per cent was available for drilling and compensating drilling losses.

At a more exotic level, some idea of process specific energy for melt penetration by a CO_2 laser can be gained from data given by Clark *et al.* (1973), who obtained specific energy consumptions for linear cutting of 6 x 10^4 lbf/in.^{2} (414 MN/m^2), or a melting efficiency of 76 per cent.

Measured Specific Energy for Liquid Jet Penetration

Hypervelocity water jets have inherently high specific energy consumption, and they would therefore normally be used in such a way that some material is left uncut by the jet itself, i.e. the kerf-and-rib technique would probably be employed. However, for planning purposes it is useful to know the basic specific energy consumption for slot-cutting.

Experimental work on the cutting of ice with high pressure water jets has been summarized by Mellor (1974), and the most recent data have been reported by Harris *et al.* (1974). Reporting of specific energy has previously been avoided because of the complications raised by secondary melting of the test slots, and by surface spalling at very small penetrations. However, under low ambient temperatures and conditions of high traverse speed and relatively low flow rate (high pressure), it appears that initial slot width is about 2.5 times the nozzle diameter, as is generally the case for deep slotting in rocks. When this width is taken for calculation of specific energy, the calculated values are maximized. Some examples of upper limit values of process specific energy are given in Fig. 8, and it can be seen that the values for low-power nozzles are very high compared with any other cutting concept.

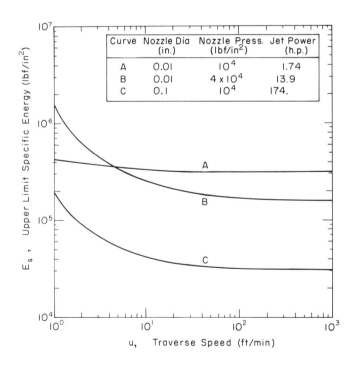

Figure 8. Examples of upper limit values for process specific energy in jet-cutting of ice.

100

The interesting feature about jets is that they permit development of tremendously high power densities. Power density (which for a given fluid and given nozzle design is proportional to nozzle pressure raised to the power 1.5) is 2.2 x 10^4 hp/in.^{2} (2.6 x 10^4 MW/m^2) for a pressure of 10^4 lbf/in.^{2} (69 MN/m^2), and 7 x 10^5 hp/in.^{2} (8.1 x 10^5 MW/m^2) for a nozzle pressure of 10^5 lbf/in.^{2} (690 MN/m^2).

Energetics of Indentation and Normal Impact

Drills that work by normal indentation or normal impact include roller rock bits, which have a static force reaction, and percussive tools that rely largely on inertial forces. Because their special characteristics are well adapted to work in strong and brittle rocks, they have not found much application in ice or fine-grained frozen soils, although they are a natural choice for drilling frozen gravels. However, there has been some interest in drilling ice and frozen fine-grained soils with vibratory tools, which can be regarded as percussive drills working at high frequency and low amplitude.

Percussive drills cover a broad spectrum, but in practice there tends to be an inverse relation between frequency and blow amplitude, since the product of frequency and blow energy gives the output power, which ordinarily stays within a limited practical range. For convenience in rough classification, percussive devices can be grouped into: (i) low frequency machines such as piling or casing hammers (powered by steam, compressed air, or internal combustion); (ii) mid-frequency machines such as percussive rock drills or impact breakers (powered by hydraulics, compressed air, or direct mechanical systems); and (iii) high-frequency machines such as "sonic" drills and pile drivers (having primary excitation by rotating eccentric mass or electromagnetic driver, sometimes with hydraulic transfer medium). For machines with moderately high power output (say 18 hp), low frequency might be represented as of the order of 1 Hz with 10^4 ft-lbf (1.4 x 10^4 J) blow energy, mid-frequency would be approaching 10 Hz with blow energy of 10^3 ft-lbf (1.4 x 10^3 J) or more, and high frequency would be 100 Hz or more with blow energy of 10^2 ft-lbf (1.4 x 10^2 J) or less.

The specific energy for indentation can vary greatly, being affected by "indexing" (spacing between individual indentations), depth of penetration (relative to indenter dimensions), and other factors. Laboratory data for low-speed (3 to 40 ft/sec, or 1 to 12 m/sec) indentation (Fig. 4) give values of 70 to 500 lbf/in.^{2} (0.5 to 3.5 MN/m^2) for ice and 600 to 2000 lbf/in.^{2} (4 to 14 MN/m^2) for frozen fine-grained soils. Results obtained from impact of high-speed inert projectiles, ranging from bullets striking at up to 4000 ft/sec (1200 m/sec) to bombs striking at up to 1000 ft/sec (300 m/sec), indicate specific energy values in the range 350 to 3500 lfb/in.^{2} (2.4 to 24 MN/m^2) (Mellor, 1972b). This somewhat indirect evidence tends to suggest that there is not much benefit to be gained by high-speed indentation once the speed is high enough to induce a brittle response. Actual percussive drilling values for specific energy are not available, but rough estimates made from measured penetration rates in ice and frozen soil suggest that they are likely to be unfavorably high. Some measurements are planned for the near future.

ROTARY DRILLING SYSTEMS

Torque and Axial Force in Rotary Systems

In a conventional rotary drilling system the power used for penetration has to be transmitted as torque and thrust in the drill string, while in a rotary system with downhole drive the corresponding torque has to be resisted by reaction "skates" and the corresponding thrust has to be provided by the weight of the unit or resisted by thrust reaction pads. Thus, while the power requirement for penetration may be inconsequential from the standpoint of energy supply, limitation of specific energy may be important in reducing the torque and thrust demands in a lightweight drill system.

With drag bits that are sharp and aggressive (high relief angle, strong positive rake), axial thrust requirements are not high in ice and fine-grained frozen soils. From personal experience the writers have found that in ice the axial thrust divided by the total width of active cutters is typically in the range 10 to 25 lbf/in. (1.8 to 4.4 N/mm) when aggressive cutters are working well; values sometimes go up to about 45 lbf/in. (8 N/mm), and down to as low as 5 lbf/in. (0.9 N/mm). In frozen fine-grained soils the values do not seem to be much higher with freshly sharpened carbides, but they increase considerably as the cutters become blunted by abrasion. The low thrust requirements for ice are easily met, even in lightweight drills, and in some cases it may be necessary to "hold back" the drill, either by keeping the drill string in tension or by limiting cutter penetration (preferably by control of effective relief angle). The electromechanical downhole ice drills that utilize the cutting head of the original CRREL corer provide far more weight than is needed for the 1.3 in. (33 mm) of active cutting edge.

With small values of axial thrust, the product of axial thrust and penetration rate represents only a small amount of power, e.g. 70 lbf (311 N) thrust at a penetration rate of 10 ft/min (3.05 m/sec) represents about 0.02 hp (0.015 kW). Thus thrust power can often be neglected in relation to torque power, and torque can be expressed conveniently in terms of specific energy.

Since torque is power divided by angular frequency, and power can be expressed as specific energy multiplied by volumetric cutting rate, torque T can be written in terms of specific energy E_s, penetration rate R, hole diameter D, and revolutions per unit time N:

$$T = \frac{R\,D^2}{8\,N}\,E_s \qquad\qquad \textbf{Eq. (7)}$$

This is for plain drilling; for coring the torque is reduced by a factor $[\,1 - (D_o/D_i)^2\,]$, where D_o and D_i are outer and inner diameters of the coring head, respectively.

From Eq. (7) it can be seen that torque is directly proportional to specific energy, and some representative values are shown graphically in Fig. 9. Torque can be reduced under some circumstances by increasing the rotational speed, but for a given power level there are limits to this effect, since chipping depth has to decrease as rotational speed increases and specific energy rises as a consequence.

Characteristics of Commercial Rotary Drills

An important aspect of systematic design procedure involves analysis of existing equipment that has evolved through practical experience to satisfy industrial needs. The first goal is to organize readily available information on commercial units in such a way that some general rules-of-thumb can be developed. In order to illustrate the procedure, we have taken some data for drag-bit auger drills; similar procedures can be followed for other classes of rotary equipment. The auger drills and large diggers provide the most direct information on power required for penetration of soil, ice, weak rock, and frozen ground, since there are no requirements for fluid or air circulation, and hoisting requirements are usually not as great as in other systems because of more limited penetration depth.

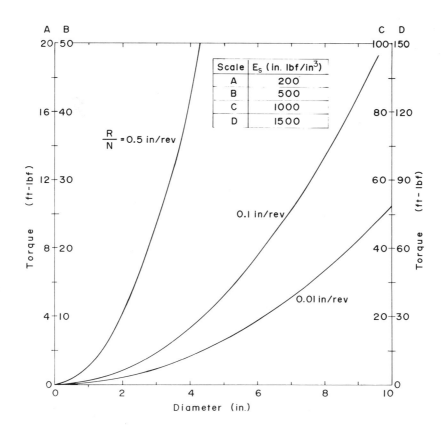

Figure 9. Required torque as a function of diameter specific energy, penetration rate, and rotary speed.

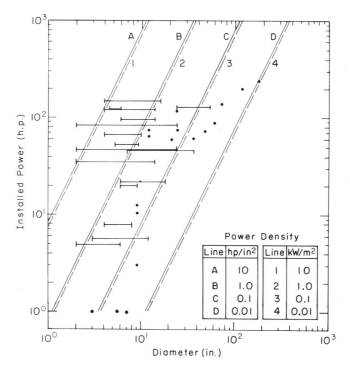

Figure 10. Installed power of existing rotary diggers and auger drills plotted against bit diameter. Lines representing a range of power density levels are superimposed.

In Fig. 10 the installed power of various augers has been plotted against bit diameter, using logarithmic scales to cover the wide size range. The assumption is that installed power is used largely for cutting and clearing in equipment of this type, so that there should be a significant dependence on diameter. From the simple mechanics of the operation, proportionality between power and the square of diameter is to be expected, i.e. any regression line drawn through the data of Fig. 10 might be expected to have a slope of 2. Actually, the plotted data cannot be expected to define any unique relation, since commercial drills of this type have to cover a range of bit sizes with a single power unit, they have to operate in a variety of material types from soils to weak rock, and they have to accept different performance limitations in terms of penetration rate and depth capability. The diameter data for some of the drills were plotted to indicate the diameter range suggested by the manufacturer, while only the largest working diameters were plotted for some of the large diggers. We have therefore drawn a set of lines that represent different power density levels, and it can be seen that the pieces of equipment represented in the plot have power densities ranging from about 0.01 hp/in.^{2} (0.01 kW/m^2) to over 10 hp/in.^{2} (10 kW/m^2). Equipment at the low end of the power density range might include very large

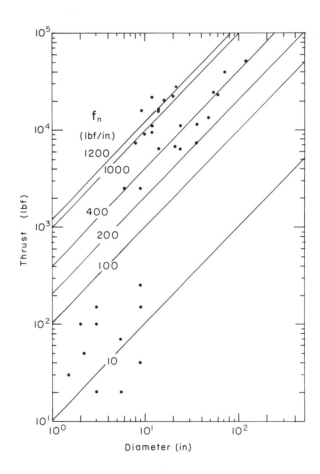

Figure 11. Maximum rated thrust plotted against maximum rated bit diameter for some existing auger drills and rotary diggers. Superimposed lines give thrust divided by diameter; these values give a measure of the normal component of cutting force where total cutter width equals bit diameter. Values can be adjusted by a factor in the range 0.4 to 1.2 in order to account for varying bit design.

104

augers that penetrate slowly and do little continuous clearing (e.g. in sinking caisson shafts), or augers designed to work only to shallow depths in very weak material (e.g. fishermen's ice augers). The high end of the power density range tends to represent large or powerful machines operating with the smallest bits that can be fitted. An interesting feature of the plot is that relatively powerful augers operate at power densities of the order of 1 kW/m^2, whereas electrothermal drills for ice and rock operate at power densities of the order of 1 MW/m^2.

In Fig. 11 rated thrust has been plotted against largest working diameter. If it is assumed that the total width of cutter edges on the bit is some simple multiple of the diameter (total width of cutter edge equals the diameter in the typical situation where the tools give 100 per cent coverage of the face), then a linear relation between thrust and diameter is expected. In real life the total cutter width may vary from 0.4 D to 1.2 D. On Fig. 11 we have drawn a set of lines that represent mean vertical thrust on unit width of the cutting tools, neglecting for present purposes the end effects of overbreak. The range is from about 200 lbf/in. (35 kN/m) to 1200 lbf/in. (210 kN/m) when total cutter width equals diameter. In laboratory cutting experiments on sedimentary rocks, the normal component of cutting force for unworn chisel-edge drag bits is typically about 200 to 300 lbf/in. (35 to 53 kN/m) for deep (but realistic) chipping. However, the normal component of cutting force increases with bit wear, in proportion to the area of the wear flat that develops on the relief face of the cutter.

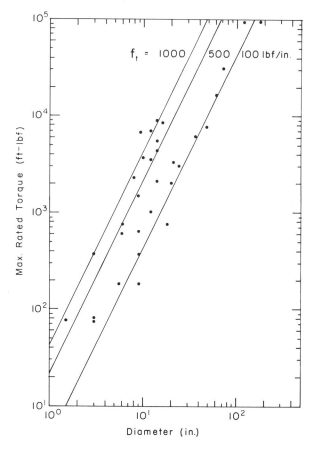

Figure 12. Maximum rated torque plotted against maximum rated bit diameter. Superimposed lines give the force of the torque divided by the diameter; these values give a measure of the tangential component of cutting force where total cutter width equals bit diameter. Values can be adjusted by a factor in the range 0.4 to 1.2 in order to account for varying bit design.

105

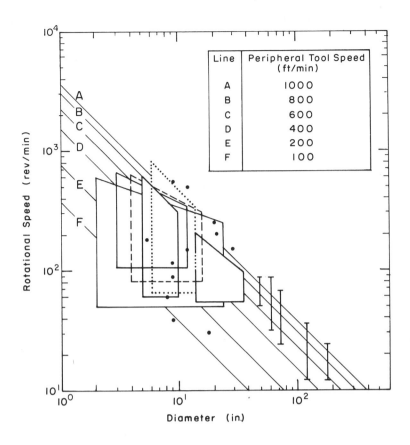

Figure 13. Rotary speed plotted against bit diameter. Superimposed lines represent various levels of peripheral tool speed.

In Fig. 12 rated torque is plotted against largest working diameters. We make the assumption that developed torque reflects the tangential component of cutting force for uniformly loaded tools, and lines have been drawn to represent various force levels when total cutter width equals bit diameter (as in the previous figure total cutter width may vary from 0.4 D to 1.2 D). The range covered by the machine data is from approximately 100 lbf/in. (17.5 kN/m) to over 1000 lbf/in. (175 kN/m). In laboratory cutting experiments on sedimentary rocks, the tangential component of cutting force for unworn chisel-edge drag bits taking cuts between 1 and 10 mm deep typically lies in the range 100 lbf/in. (17.5 kN/m) (for shallow cuts or for sharp tools with strong positive rake) to over 1000 lbf/in. (175 kN/m) (for tools taking deep cuts). The tangential force component tends to be less dependent on wear than the normal component, especially with negative-rake tools.

In Fig. 13 rotational speed has been plotted against bit diameter, with the intention of defining the linear velocity of the peripheral tools, i.e. the maximum tool speed. However, some caution is called for in preparing and interpreting such a graph, since a drill that has a range of bit sizes and rotational speeds does not necessarily have the capability of effectively using the largest bits at the highest speeds because of torque or power limitations. For this reason, only rpm values that appeared most reasonable were plotted for the various diameters. In broad terms, the maximum potential tool speeds indicated by the graph are in the range 100 to 1000 ft/min

106

(0.51 to 5.1 m/sec), which is the range normally considered to be optimum in the design of drag-bit mining tools (wear becomes unacceptably high at greater speeds in abrasive rock).

CONCLUSION

While many drilling systems are bewilderingly complex at first sight, they provide only three simple basic functions: penetration, material removal from the hole, and hole stabilization. There are many ways of meeting each of these functional requirements, but because of the need for some degree of compatibility between each functional element of the system, the number of practical combinations is limited.

The minimum power required to meet a given performance specification can be estimated for each functional element from simple physical considerations, provided that certain material properties for the ground material are known. These power estimates are useful for comparing concepts and assessing compatibility of the individual elements. They also provide a basis for estimating torque and axial force in rotary systems.

Field data for drilling devices operating in ice and frozen soils show wide discrepancies in performance, and suggest that many past operations have fallen far short of attainable energetic efficiency levels.

Some drilling concepts are inherently less efficient than competing concepts in energetic terms, but may still be attractive because they offer easy transmission of energy, possibly coupled with a potential for high power density at the drill tip. Practical limitations on power density can set a limit to potential penetration rate for some drilling concepts.

New drilling units for unusual ground conditions sometimes evolve unsystematically through successive empirical adaptations and modifications of components that are marginally suitable or weakly compatible. However, it now seems possible to reduce the dependence on empiricism in new development, since the data and methodology for an analytical approach are becoming available. This is particularly true in the case of rotary drilling, where current research into the kinematics, dynamics and energetics of rotary cutting is yielding systematic data on penetration and chip removal.

REFERENCES

Aamot, H.W.C., 1967a, Heat transfer and performance analysis of a thermal probe for glaciers: U.S. Army CRREL Technical Report 194.

Aamot, H.W.C., 1967b, The Philberth probe for investigating polar ice caps: U.S. Army CRREL Special Report 119.

Aamot, H.W.C., 1968, A buoyancy-stabilized hot-point drill for glacier studies: *Journal of Glaciology,* v. 7, no. 51, pp. 493-498.

Abel, J.F., 1960, Permafrost tunnel, Camp Tuto, Greenland: U.S. Army SIPRE Technical Report 73.

Abel, J.F. 1961, Under-ice mining techniques: U.S. Army CRREL Technical Report 72.

Arctic Construction and Frost Effects Laboratory, U.S. Army (USA ACFEL), 1954, Development of power ice coring rig, U.S. Army ACFEL Technical Report 46.

Armstrong, P.E., 1974, Subterrene electrical heater design and morphology: Los Alamos Scientific Laboratory Informal Report LA-5211-MS, University of California, Los Alamos, New Mexico.

Bailey, J.J., 1967, A laboratory study of the specific energy of disengagement of frozen soils: U.S. Army CRREL Internal Report 99 (unpublished), conducted by Creare, Inc., Hanover, New Hampshire.

Barker, J.S., 1964, A laboratory investigation of rock cutting using large picks: *International Journal of Rock Mechanics and Mining Sciences,* v. 1, no. 4, pp. 519-534.

Beard, W.H. and C.R. Hoffman, 1967, Polar construction equipment—Investigation of drilling equipment: U.S. Naval Civil Engineering Laboratory Technical Note N-887.

Browning, J.A. and E.M. Fitzgerald, 1964, Internal burners for the drilling and slotting of permafrost: U.S. Army CRREL Internal Report 98 (unpublished).

Browning, J.A. and J.F. Ordway, 1963, The use of internal burners for the working of permafrost and ice: U.S. Army CRREL Internal Report 97 (unpublished).

Bruce, W. and R. Morrell, 1969, Principles of rock cutting applied to mechanical boring machines: Proceedings of the 2nd Symposium on Rapid Excavation, Sacramento, California.

Cherepanov, N.V., 1969, A new annular ice auger: *Problems of the Arctic and Antarctic,* no. 32, pp. 122-125 (Russian) (National Science Foundation Translation, 1970, pp. 506-510).

Clark, A.F., J.C. Moulder and R.P. Reed, 1973, Ability of a CO_2 laser to assist icebreakers: *Applied Optics,* v. 12, no. 6, pp. 1103-1104.

Gaye, F., 1972, Efficient excavation with particular reference to cutting head design of hard rock tunnelling machines: *Tunnels and Tunneling,* Jan/Feb, Mar/Apr.

Harris, H.D., M. Mellor and W.H. Brierley, 1974, Jet cutting tests on floating ice: Laboratory Memorandum no. GS-216, Gas Dynamics Laboratory, Division of Mechanical Engineering, National Research Council of Canada.

Hodge, S.M., 1971, A new version of a steam-operated ice drill: *Journal of Glaciology,* v. 10, no. 60, pp. 387-393.

Hoffman, C.R. and E.H. Moser, 1967, Polar construction equipment—Drilling tests in ice and ice-rock conglomerate: U.S. Naval Civil Engineering Laboratory Technical Note N-937.

Howorka, F., 1965, A steam-operated ice drill for the installation of ablation stakes on glaciers: *Journal of Glaciology,* v. 5, no. 41, pp. 749-750.

Huck, R.W., 1969, Vibratory pile driving and coring in permafrost: U.S. Army CRREL Technical Note (unpublished).

Kovacs, A., 1970, Augering in sea ice: U.S. Army CRREL Technical Note (unpublished).

Kovacs, A., 1974, Ice augers (Continuous flight, lightweight, man-portable): U.S. Army CRREL Technical Note (unpublished).

Kovacs, A., M. Mellor and P.V. Sellmann, 1973, Drilling experiments in ice: U.S. Army CRREL Technical Note (unpublished).

LaChapelle, E., 1963, A simple thermal ice drill: *Journal of Glaciology,* v. 4, no. 35, pp. 637-642.

Lange, G.R., 1964, Additional data on effectiveness of various means of cutting frozen soil, rock and ice: U.S. Army CRREL Technical Note (unpublished).

Lange, G.R., 1968, Rotary drilling in permafrost: U.S. Army CRREL Technical Report 95.

Lange, G.R., 1973, Construction of an unattended seismological observatory (USO) in permafrost: U.S. Army CRREL Special Report 113.

Lundquist, R.G., 1968, Rock drilling characteristics of hemispherical insert bits: Thesis, University of Minnesota.

McAnerney, J.M., 1968, Tunneling in a subfreezing environment: University of Minnesota Tunnel and Shaft Conference, Minneapolis.

McCoy, J.E., 1960, Performance of a Williams auger in permafrost: U.S. Army SIPRE Special Report 38.

Mellor, M., 1971, Blasting tests in frozen ground, 1971: U.S. Army CRREL Technical Note (unpublished).

Mellor, M., 1972a, Normalization of specific energy values: *International Journal of Rock Mechanics and Mining Sciences,* v. 9, no. 5, pp. 661-663.

Mellor, M., 1972b, Use of impact tools for penetrating and excavating frozen ground: U.S. Army CRREL Technical Note (unpublished).

Mellor, M., 1972c, Design parameters for a rotary excavating attachment: U.S. Army CRREL Technical Note (unpublished).

Mellor, M., 1974, Cutting ice with continuous jets: Second International Symposium on Jet Cutting Technology, Cambridge.

Mellor, M., 1975, Cutting frozen ground with disc saws: U.S. Army CRREL Technical Report 261.

Mellor, M. and I. Hawkes, 1972, Hard rock tunneling machine characteristics: Rapid Excavation and Tunneling Conference, Chicago.

Mellor, M. and P.V. Sellmann, 1970, Experimental blasting in frozen ground: U.S. Army CRREL Special Report 153.

Mellor, M., P.V. Sellmann and A. Kovacs, 1973, Drill penetration rates for ice and frozen ground: U.S. Army CRREL Technical Note (unpublished).

Miller, M.H. and D.L. Sikarskie, 1968, On the penetration of rock by three-dimensional indentors: *International Journal of Rock Mechanics and Mining Sciences,* v. 5, no. 5, pp. 375-398.

Murphy, D.J. and R.G. Gido, 1973, Heat loss calculations for small diameter subterrene penetrators: Los Alamos Scientific Laboratory Informal Report LA-5207-MS, University of California, Los Alamos, New Mexico.

Nikolaev, A.F. and E.A. Trubina, 1969, Investigation of the drilling process in ice: *Rybnoe Khoziaistvo,* no. 6, pp. 52-53.

Peng, T., 1958, The investigation of ice cutting process: U.S. Army SIPRE Internal Report 87 (unpublished).

Rad, P.F., 1970, Effects of laser radiation on some cutting characteristics of granite: Ph.D. Thesis, Massachusetts Institute of Technology.

Rausch, D.O., 1958, Ice tunnel, Tuto area, Greenland: U.S. Army SIPRE Technical Report 44.

Roxborough, F.F., 1973, Cutting rock with picks: *The Mining Engineer,* June, pp. 445-455.

Sellmann, P.V. and M. Mellor, 1974, Man-portable drill for ice and frozen ground—Preliminary development report: U.S. Army CRREL Technical Note (unpublished).

Shreve, R.L., 1962, Theory of performance of isothermal solid-nose hot points boring in temperate ice: *Journal of Glaciology,* v. 4, no. 32, pp. 151-160.

Shreve, R.L. and R.P. Sharp, 1970, Internal deformation and thermal anomalies in lower Blue Glacier, Mount Olympus, Washington, U.S.A.: *Journal of Glaciology,* v. 9, no. 55, pp. 65-86.

Stacey, J.S., 1960, A prototype hot point for thermal boring on the Athabaska Glacier: *Journal of Glaciology,* v. 3, no. 28, pp. 783-786.

Trans-Alaska Pipeline System, 1969, Trenching equipment evaluation tests (unpublished).

Ueda, H.T. and D.E. Garfield, 1968, Drilling through the Greenland Ice Sheet: U.S. Army CRREL Special Report 126.

Ueda, H.T. and D.E. Garfield, 1969a, Core drilling through the Antarctic Ice Sheet: U.S. Army CRREL Technical Report 231.

Ueda, H.T. and D.E. Garfield, 1969b, The USA CRREL drill for thermal coring in ice: *Journal of Glaciology,* v. 8, no. 53, pp. 311-314.

Ueda, H.T. and D.E. Garfield, 1970, Deep core drilling at Byrd Station, Antarctica: *International Association of Scientific Hydrology, Publication,* no. 86, International Symposium on Antarctic Glaciological Exploration (ISAGE), Hanover, New Hampshire, 1968, pp. 53-62.

Zelenin, A.N., 1959, *Rezanie Gruntov:* Izdat. Akad. Nauk, SSSR, Moscow.

Zelenin, A.N., 1968, *Fundamentals of ground excavation by mechanical means:* Machinostroenie, Moscow (text in Russian).

THERMAL CORE DRILLING IN ICE CAPS IN ARCTIC CANADA

W.S.B. Paterson
Polar Continental Shelf Project
Department of Energy, Mines and Resources
Ottawa, K1A OE4, Canada

ABSTRACT

The CRREL shallow-hole thermal coring drill has been used to drill a 121-m borehole through the Meighen Ice Cap and three holes (230, 299 and 299 m) in the ice cap on Devon Island. Three of the four holes reached bedrock; in the 230-m hole, the drill became frozen in and was lost. Operating conditions, the performance of the drill, and problems encountered are described.

Drilling Program

The boreholes that have resulted from the Polar Continental Shelf Project drilling program in the Canadian Arctic Islands are listed in Table 1. The Meighen Ice Cap is a small (80 km^2) low-lying ice cap which is thought to have originated after the end of the post-glacial Climatic Optimum. The borehole is near the highest point of the ice cap which is also the region of maximum ice thickness. Accumulation is normally in the form of superimposed ice so that drilling was in ice throughout. The ice cap on Devon Island has an area of about $15,500$ km^2 and consists of an east-west ridge drained by valley glaciers, most of which extend to sea level. The three boreholes are within 300 m of each other, near the summit ridge, and about 7 km west of its highest point. The summit ridge overlies a region of high bedrock so that the ice at the boreholes is relatively thin. Nevertheless, it spans a time period of about 100,000 years. The top 60 m of each borehole is in firn. The ice is well below the pressure-melting temperature in all cases.

The borehole measurements and the analyses in progress on the Devon cores are listed in Table 2. Two boreholes to bedrock were required because the down-borehole extraction of CO_2 for radiocarbon dating leaves the hole blocked by refrozen meltwater. Thus another borehole was needed for flow and temperature measurements. The third hole is blocked by the drill which became frozen in at 230 m. These boreholes have provided an opportunity to test the consistency of the results of oxygen isotope and other analyses between adjacent boreholes. The majority of the analyses listed in Table 2 were also performed on the Meighen core. In addition, closure rate and temperature were measured in the borehole.

Drilling Techniques

The CRREL thermal coring drill, described by Ueda and Garfield (1969), was used. Of the

Table 1

Details of Boreholes

Ice Cap	Latitude, °N	Longitude, °W	Date	Depth, m	Time Taken, days	Bedrock Reached	Ice Temperature, °C
Meighen	79.9	99.1	June 1965	121	22	yes	-15 to -19
Devon	75.3	82.3	May 1971	230	16	no	-20 to -23
Devon	75.3	82.3	May 1972	299	19	yes	-18 to -23
Devon	75.3	82.3	May 1973	299	16	yes	-18 to -23

Table 2

Devon Island Boreholes: Studies in Progress

Core analyses

1. Count of particulates — Polar Shelf Project
2. Electrolytic conductivity — Polar Shelf Project
3. Ice-fabric analysis — Polar Shelf Project
4. Oxygen isotopes — University of Copenhagen
5. Pollen — Geological Survey of Canada
6. Gas content — C.N.R.S., Grenoble
7. C^{14} dating — University of Bern
8. Si^{32} dating — University of Copenhagen

Borehole measurements

9. Change of inclination — Polar Shelf Project
10. Closure rate — Polar Shelf Project
11. Vertical strain rate — Polar Shelf Project
12. Temperature — Polar Shelf Project
13. Measurement of velocity of radio waves — Scott Polar Research Institute

ancillary equipment, the tower was built by CRREL while the hoist was designed and constructed by the Canadian Longyear Company. An Onan 5-kW gasoline-driven generator provides power for the drill and hoist. The equipment weighs about 1400 kg; the heaviest piece (the drum and 450 m of cable) weighs 450 kg. The equipment is transported to the site by a Twin Otter aircraft fitted with skis. For the first drilling, there was little shelter around the rig; as a result, bad weather caused frequent interruptions. Subsequently, all the equipment has been mounted inside a Parcoll building, with part of one roof section removed to make room for the tower. Drilling can then proceed irrespective of the weather, except in very high winds. The hoist and tower are placed on a substantial wooden platform set on the snow surface; it has never been necessary to relevel the platform during any drilling. The uppermost 3 or 4 m of each borehole is cased with plastic pipe to keep out surface meltwater. A cap is kept on the casing except when the drill is being raised or lowered through the top of it.

The CRREL drill produces a hole of diameter 16.4 cm and a core of diameter 12.2 cm in 1.5-m sections. About 50 minutes are required to drill each section and about 15 liters of meltwater are produced. The water is pumped continuously from the drill head into a storage tank in the upper part of the drill. We preserve the meltwater, filter it for pollen, and perform chemical analyses on it. We have used two versions of the drill. The main improvements in the later model were an improved mechanism for suspending the drill from the cable, a better pump motor, the inclusion of a water-level indicator in the storage tank, and the elimination of valves in the tubes which carry the meltwater from the drill head to the tank. These modifications eliminated several problems encountered with the first drill and the new version proved much more reliable, as shown by reduced drilling times for the 1972 and 1973 boreholes (Table 1).

Core recovery from the boreholes was complete but many of the cores were badly fractured. On Devon Island, starting at a depth of about 70 m, the cores had spoon-shaped fractures at the places where they were gripped by the core catchers. As depth increased, the pattern gradually changed to horizontal fractures; the transformation was complete by about 160 m, at which point there were horizontal fractures throughout each core. Horizontal fracturing continued to within a few meters of the base of the ice. In the lowest few meters, sections of core with spoon-shaped fractures alternated with horizontally-fractured sections. The fractures probably result from release of hydrostatic pressure, coupled with thermal shock. The change in pattern may be related to the development of a preferred orientation in the ice crystals, but this aspect of the cores has not yet been studied in detail.

Our normal practice is to drill for 14 or 15 hours per day because lack of manpower prevents round-the-clock drilling. Closure of the hole during shutdowns has never caused any problem; measured closure rates near the base of the ice are less than 0.3 mm/day. No dirt bands were encountered in the ice on Devon Island. On Meighen Ice Cap, pockets of dirt reduced drilling speeds by up to 25 per cent and also caused the valves in the suction tubes to stick open. This was a serious problem because it allowed water to escape into the hole while the drill was being raised to the surface. However, these valves have been eliminated in later models of the drill. The boreholes have shown little tendency to deviate from the vertical. The horizontal displacement between top and bottom was 0.7 and 1.0 m in the two boreholes where this was measured.

The major setback in the drilling program was the loss of the drill in the 230-m hole on Devon Island. This probably resulted from the leakage of water from the storage tank while the drill was down the hole. The tank was constructed with three small holes near the bottom and the plug in one of them had previously come out while the drill was being raised. The most likely explanation is that one of these plugs came out while the drill was down the hole. This would release about 10 liters of water, about half-way along the length of the drill, into ice at a temperature of -20°C. The alternative explanation, a pump failure, is considered less likely because the pump motor was drawing its normal current when the accident occurred. The first sign of trouble was a reduction of cable tension. Attempts to bring the drill to the surface failed because the motor on the hoist stalled repeatedly. When this was prevented by boosting the output from the generator, the main sprocket on the hoist broke. This put the hoist out of action and greatly reduced the chance of saving the drill. About 70 liters of ethylene-glycol solution were pumped down the hole, by means of a rubber hose to ensure that the solution reached the top of the drill, and an electric heater was kept running in the solution. These attempts to free the drill were unsuccessful, however, and were abandoned after three days. The cable was then cut a short distance above the drill to leave the hole free for measurements.

During the 1973 drilling, a plastic insulator became detached from the drill head and lodged between the head and the ice, reducing drilling speed to less than half normal. It was removed by the following method. The Swiss probe for melting ice down the borehole for gas extraction was run for approximately one hour. This melted a large pear-shaped cavity and it was hoped that the insulator would move to the lowest point of the cavity. This is apparently what happened because, after waiting for about 30 hours for the water to refreeze, the insulator was recovered in the next core.

An improvement we would suggest is a method of orienting the cores that does not rely on a magnetic compass. Although this is a problem peculiar to our area, which is only a few hundred kilometers from the magnetic pole, the lack of core orientation is a serious drawback. We would also suggest an improved system of indicating the water level in the storage tank. The system in our drill appears to be unreliable and in fact we have stopped using it. To avoid the possibility of overflow of the tank, the drill is run for a maximum of 55 minutes, even if the full 1.5 m has not been drilled by then. The pump was expected to be a possible weakness in the drill and so we flush it with alcohol after every run and fit a new one after about 200 or 250 m of drilling. In this way we have avoided any problems with it, except perhaps when the drill was lost.

Summary

The latest version of the CRREL drill is an efficient method of taking cores from depths of a few hundred meters. An idea we put forward for consideration is a drill that collects only meltwater, not ice. Such a drill would no doubt be smaller and lighter than the present one. It would be sufficient for most purposes because the cores have to be melted for most of the analyses performed on them.

Acknowledgment

As will be apparent, our drilling equipment and techniques are those developed by Lyle Hansen and other members of the Technical Services Division at CRREL. I should like to take this opportunity of thanking them for all their assistance.

REFERENCE

Ueda, H.T. and D.E. Garfield, 1969, The USA CRREL drill for thermal coring in ice: *Journal of Glaciology,* v. 8, no. 53, pp. 311-314.

THE THERMAL PROBE DEEP-DRILLING METHOD BY EGIG IN 1968

AT STATION JARL-JOSET, CENTRAL GREENLAND

Karl Philberth
D 8031 Puchheim/München
Peter-Rosegger-Str. 6
West Germany

ABSTRACT

A special thermal probe deep-drilling method was used by Expédition Glaciologique Internationale au Groenland (EGIG) 1968. The wire for the transmission of the electric power and the measured values pays out of the probe and becomes fixed in the refreezing meltwater.

Two probes were constructed, each with a diameter of 10.8 cm and lengths of about 2.5 m (probe II) and 3 m (probe I). With the available maximum power of 3.7 kW the velocity was 2 m/hr.

The probes had sufficient wire for the penetration of the ice sheet (2500 m), but the breakdown of the main heater stopped probe I at a depth of 218 m and probe II at a depth of 1005 m. The ice temperatures recorded after cooling occurred were: -29.0°C at 218 m, -29.3°C at 615 m and -30.0°C at 1005 m depth. The method as such worked without significant problems.

Thermal probes of this type are relatively inexpensive (about $15,000), easy to handle (about 10 cm diameter, 200-300 cm length) and work fast (50-100 m/day or more). A small summer expedition could penetrate 4000 m of ice or more.

The summer 1968 campaign of EGIG (Expédition Glaciologique Internationale au Groenland) contained a German-Swiss thermal deep-drilling program at station Jarl-Joset (33°28′ W, 71°21′ N, 2865 m above sea level), where the ice thickness is 2500 m. Two penetrations were made with a thermal probe system which I originated (K. Philberth, 1962a, 1962b, 1962c, 1966a, 1966b, 1970), and then developed together with B.L. Hansen and H. Aamot (1967a, 1967b, 1968, 1970). Probe I corresponds closely to the drawings provided by H. Aamot, but probe II is different in some respects.

The probes were produced in Germany by the German-American Eastman International Company GmbH. The field work was done by the Benedictine Father H. Jännichen (physicist), E. Gmeineder (engineer) and me, in cooperation with five French technicians.

The characteristic feature of this type of probe is that the wire for the transmission of electric power and remote sensing pays out of the advancing probe and becomes fixed in the re-freezing meltwater. The probes are thus not retrievable. It is no problem to supply sufficient wire and sufficient sidewall heating for the penetration of the thickest and coldest Antarctic ice layers. The sidewall heating can be supplied by the heat of the ohmic resistance of the stored wire.

Probe Specifications

Figure 1 shows the EGIG thermal probes I and II and the lower part of both. Probe I had an insulated and a bare wire, an oil reservoir which was attached in the field, and was 292 cm long (including reservoir) and 10.8 cm in diameter. Probe II had two insulated wires and no reservoir, and was 255 cm long and 10.8 cm in diameter.

The lower part of the probes contained the cartridge main heater, the mercury heat flux control for the stabilization of verticality (K. Philberth, 1964, 1966a), the pressure-protected box for the instrumentation and auxiliary heaters above and below it.

Figure 2 shows the upper parts of probes I and II. The sidewall around the bare-wire coil and of the oil reservoir of probe I contained a 100-ohm resistor wire for additional heating. The wire coils were "orthocyclic" (Lenders, 1962; Aamot, 1969). Each coil contained about 3150 m of copper wire, stored in 23 layers of 600 windings each (insulated wire) and 29 layers of 485 windings each (bare wire). The bare wire diameter of probe I was 0.090 cm, and the insulated wire was 0.095 cm (net) and 0.126 cm (gross). The insulation was Teflon-sealed Kapton tape.

Figures 3, 4, and 5 are the circuit diagrams of probes I and II and of the surface instrumentation for control and supervision, respectively. In the pressure-protected box are a 12-step relay, thermistors, a pair of strain gauges, a simple inclinometer, calibration resistors and diodes. Part of the surface instrumentation is the "Last-Brücke" (compare Fig. 9) for the continuous supervision of probe performance. It is a bridge designed by B. Philberth in which the complex impedance of the probe (L, R, C) is compensated, independent of the frequency, by a reciprocal network (c, r, l). The precision is in the order of 0.01 - 0.1 per cent; progress of less than 30 cm can be observed.

Performance

The cartridge heater of probe I experienced a short circuit at a depth of 218 m, that of probe II at a depth of 1005 m. The short circuit ended the run. The cause was probably moisture content of the insulation, resulting in electrolytic action and then failure.

Probe I measured a cooling curve for a 5-day period at a depth of 218 m, resulting in a virgin ice temperature T_j = -29.0°C. Probe II was stopped intentionally at a depth of 615 m, where a cooling curve for a 6.5-day period was measured, resulting in T_j = -29.3°C. At a depth of 1005 m the cooling curve of probe II was interrupted after 5.2 hours by wire breakage. T_j was determined to be -30.0°C. Figures 6 and 7 give the cooling curves. The changing power in probe I in the last 100 minutes (Fig. 6) caused an irregularity of its cooling curve in the first 300 minutes.

The depth of the probes was measured by four methods: inductance of the coil, time integration of the velocity, resistance of the length of the bare wire (probe I only), and hydrostatic pressure measured by strain gauges. The first two methods worked continuously. The inductance method made use of the harmonics (100, 300 cps) of the DC power supply: the values of c, r_w (AC resistance) and r_G (DC resistance) are read on the "Last-Brücke." They depend slightly on the value of l and are functions of the depth of the probe. Figure 8 shows these relations for the first section (50-615 m) of probe II, and Fig. 9 for the second section (615-1005 m).

The inclination of probe I for its total path and of probe II for the second section was always less than 4°. However in the first section probe II reached an inclination up to 10° (Fig. 8), probably because a piece of polyethylene had fallen into the hole and adhered to the sidewall of the probe.

Probe I (Fig. 10) was started at a depth of 7 m. Down to a depth of 35 m the firn absorbed the meltwater. Therefore the mean coil temperature T_{Wickel} (T_{coil}) reached 90^OC for a current J of 1.8 A. Later, for J equal to 2.7 A (corresponding to a total power N_{total} of 3.7 kW) and a velocity of 2.0 m/hr, the temperatures were as follows: less than 80^OC (mean value) for the coil, less than 55^OC for the thermistors, and about 300^OC for the Chromax wire in the cartridge heater. Probe II (Fig. 11) was started at a depth of 45 m and reached a current J of 2.3 A (corresponding to N_{total} = 3.7 kW) and a velocity of 1.9 m/hr. Under these conditions the temperatures were as follows: less than 80^OC (mean value) for the coils, less than 50^OC for the thermistors, and less than 250^OC for the Chromax wire. The level of the heavy interspace oil decreased quicker than expected.

The reheating process of probe II at a depth of 615 m is shown in Fig. 12. For about 30 hours only the coils were heated (Position 12); the heating current was increased until the temperature T_{Wickel} of the coils made it evident that the upper part of the probe had reached the melting point.

Glaciological Results

Extrapolation (K. Philberth, 1972; Philberth and Federer, 1971, 1973, 1974) of the measured temperatures (see above) yields a bottom temperature at this location of -12^OC. The difference between measured and calculated temperature profiles down to a depth of 600 m is about the same as the corresponding difference at Camp Century and can be explained by short-period changes of the climate. The Ice Age temperature is likely to have been by about 6^OC lower than today. The glaciological conditions would be favorable for the disposal of radioactive waste (B. Philberth, 1956, 1961; K. Philberth, 1976a, 1976b).

Suggestions

Low-voltage heaters are more reliable than high-voltage heaters. The probe could be equipped with a semiconductor DC/AC-converter and a small transformer, changing the high voltage DC power supply into low voltage AC with e.g. 1000 cps. Instead of the heavy oil an antifreeze solution could be applied, which is trapped in the probe and gets mixed with the penetrating meltwater. The destructive effect of unmixed meltwater inside the coil could be reduced by a very small auxiliary heater at the top. There is no meltwater penetration problem at all if the coil (with coaxial or double wire) is stored outside the (slender) cylinder of the probe; the wire of the coil could be stabilized by a semi-adhesive surface and taken off from "over-head."

For the stabilization of the verticality various other methods could be applied: controlled on-off-switching of heaters, flat bottom (K. Philberth, 1964, 1966a), pendulum method (Aamot, 1970), eccentric gravity center of the probe with more bottom heating at the side where the center of gravity is located, sharp fins at the upper end of the probe (B. Philberth).

Instead of two wires, one coaxial or multiaxial cable could be applied, if this is sufficiently reliable. After accidental wire break or wire short-circuit (coaxial cable), at least temperature measurements should be guaranteed for some days, e.g. by wireless or by "semi-wireless" transmission. The latter means transmission by frequency-modulated pulses, which can jump over wire interruption by virtue of its capacity.

A detailed report on the EGIG drilling and suggestions for improvement of probes is in press (K. Philberth, in press).

Acknowledgments

Excellent advice and cooperation were provided by B.L. Hansen and H. Aamot of the U.S. Army Cold Regions Research and Engineering Laboratory. The work was supported by the Deutsche Forschungsgemeinschaft and by the Swiss Federal Institute for Snow and Avalanche Research.

REFERENCES

Aamot, H.W.C., 1967a, The Philberth probe for investigating polar ice caps: U.S. Army CRREL Special Report 119.

Aamot, H.W.C., 1967b, Heat transfer and performance analysis of a thermal probe for glaciers: U.S. Army CRREL Technical Report 194.

Aamot, H.W.C., 1968, Instrumented probes for deep glacial investigations: *Journal of Glaciology*, v. 7, no. 50, pp. 321-328.

Aamot, H.W.C., 1969, Winding long, slender coils by the orthocyclic method: U.S. Army CRREL Special Report 128.

Aamot, H.W.C., 1970, Development of a vertically stabilized thermal probe for studies in and below ice sheets: *Journal of Engineering for Industry*, v. 92, ser. B, no. 2, pp. 263-268 (Transactions of the ASME, Paper No. 69-WA/UnT-3).

Lenders, W.L.L., 1962, Das orthozyklische Wickeln von Spulen: *Philips' Technische Rundschau*, Jahrg. 23, Nr. 12, pp. 401-416 (reprints in English language available).

Philberth, B., 1956, Beseitigung radioaktiver Abfallsubstanzen: *Atomkern-Energie*, I. Jahrg., Heft 11/12, pp. 396-400.

Philberth, B., 1961, Beseitigung radioaktiver Abfallsubstanzen in den Eiskappen der Erde: *Schweizerische Zeitschrift für Hydrologie*, Vol. XXIII, Fasc. 1, pp. 263-284.

Philberth, K., 1962a, Une méthode pour mesurer les températures à l'intérieur d'un Inlandsis: *Comptes Rendus des Séances de l'Académie des Sciences*, t. 254, no. 22, pp. 3881-3883.

Philberth, K., 1962b, Remarque sur une sonde thermique pour mesurer la temperature des couches de glace: *Comptes Rendus des Séances de l'Académie des Sciences,* t. 255, pp. 3022-3024.

Philberth, K., 1962c, Ecoulement de la glace Groenlandaise: *Revue de Géomorphologie Dynamique,* No. 1, 2, 3.

Philberth, K., 1964, Uber zwei Elektro-Schmelzsonden mit Vertikal-Stabilisierung: *Polarforschung,* Jahrg. 34, Heft 1/2, pp. 278-280.

Philberth, K., 1966a, Sur la stabilisation de la course d'une sonde thermique: *Comptes Rendus des Séances de l'Académie des Sciences,* t. 262, pp. 456-459.

Philberth, K., 1966b, Eine Schmelzsonde zur Messung des Temperaturprofils in Eiskalotten: *Umschau in Wissenschaft und Technik,* Heft 11, p. 360.

Philberth, K., 1970, Thermische Tiefbohrung in Zentralgrönland: *Umschau in Wissenschaft und Technik,* Heft 16, pp. 515-516.

Philberth, K., 1972, Factors influencing deep ice temperatures: *Nature, Physical Science,* v. 237, no. 72, pp. 44-45.

Philberth, K., 1976a, On the temperature response in ice sheets to radioactive waste deposits: *Journal of Glaciology,* v. 16, no. 74.

Philberth, K., 1976b, Future regard to the atomic waste disposal problem: *Journal of Glaciology,* v. 16, no. 74.

Philberth, K., in press, Die thermische Tiefbohrung in Station Jarl-Joset und ihre theoretische Auswertung: *Meddelelser om Grønland.*

Philberth, K. and B. Federer, 1971, On the temperature profile and the age profile in the central part of cold ice sheets: *Journal of Glaciology,* v. 10, no. 58, pp. 3-14.

Philberth, K. and B. Federer, 1973, On the temperature gradient in cold ice sheets: Interner Bericht Nr. 530 des Eidgen. Institut für Schnee- und Lawinenforschung, Davos (Schweiz).

Philberth, K. and B. Federer, 1974, On the temperature gradient in the upper part of cold ice sheets: *Journal of Glaciology,* v. 13, no. 67, pp. 148-151.

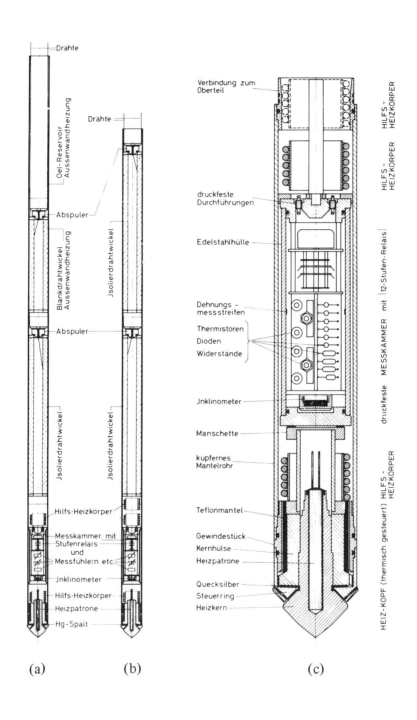

Figure 1. Thermal probes: (a) I; (b) II; (c) lower part of I and II.

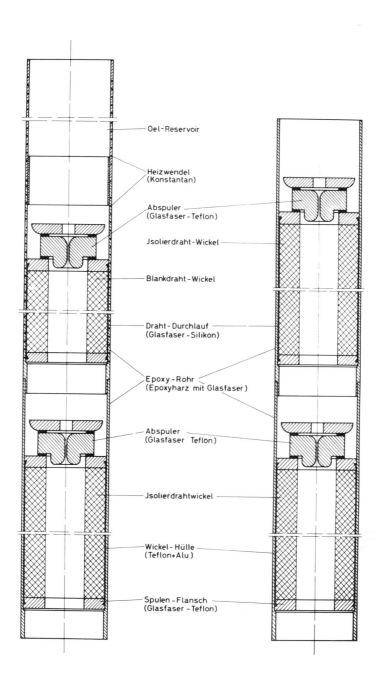

Oel-Reservoir

Heizwendel
(Konstantan)

Abspuler
(Glasfaser-Teflon)

Jsolierdraht-Wickel

Blankdraht-Wickel

Draht-Durchlauf
(Glasfaser-Silikon)

Epoxy-Rohr
(Epoxyharz mit Glasfaser)

Abspuler
(Glasfaser Teflon)

Jsolierdrahtwickel

Wickel-Hülle
(Teflon+Alu)

Spulen-Flansch
(Glasfaser-Teflon)

Figure 2. Upper parts of probes I (left) and II (right).

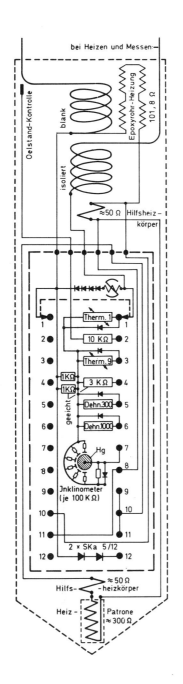

Figure 3. Circuit diagram for probe I.

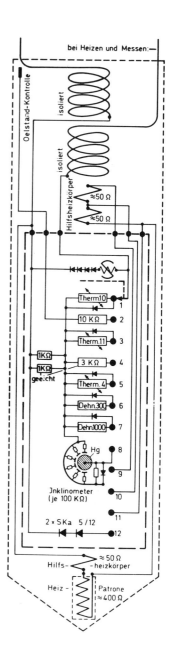

Figure 4. Circuit diagram for probe II.

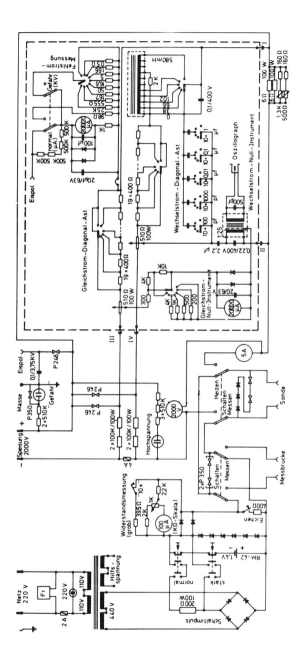

Figure 5. Surface instrumentation for control and supervision of the system.

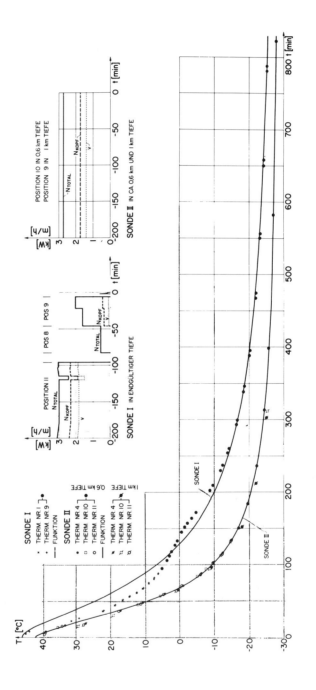

Figure 6. Cooling curves for probes I and II.

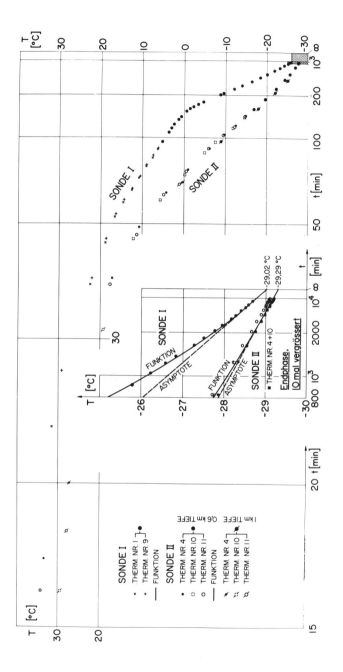

Figure 7. Cooling curves for probes I and II.

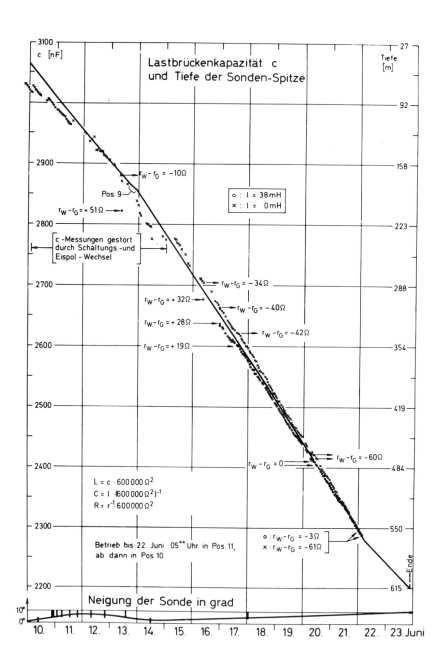

Figure 8. Relationship of wire inductance L and resistance R to depth for probe II (50-615 m).

128

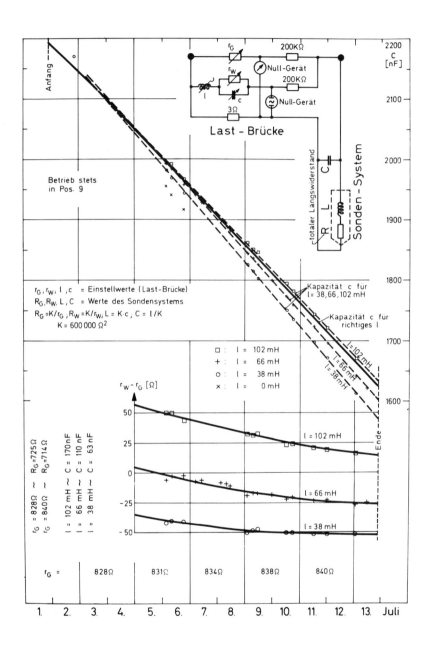

Figure 9. Relationship of wire inductance L and resistance R to depth for probe II (615-1005 m).

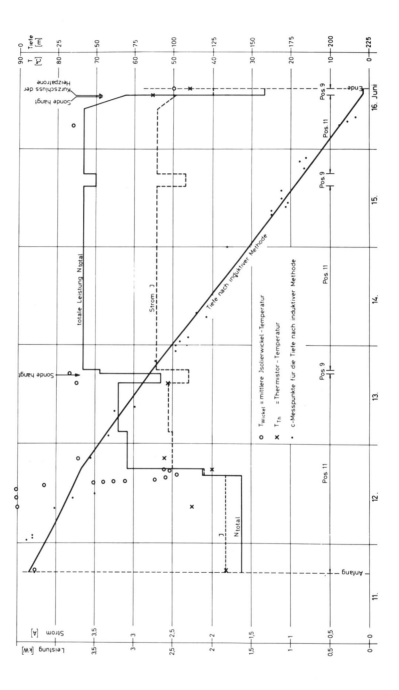

Figure 10. Total power, current and temperatures of coil (mean) and thermistors for probe I between surface and 218 m.

130

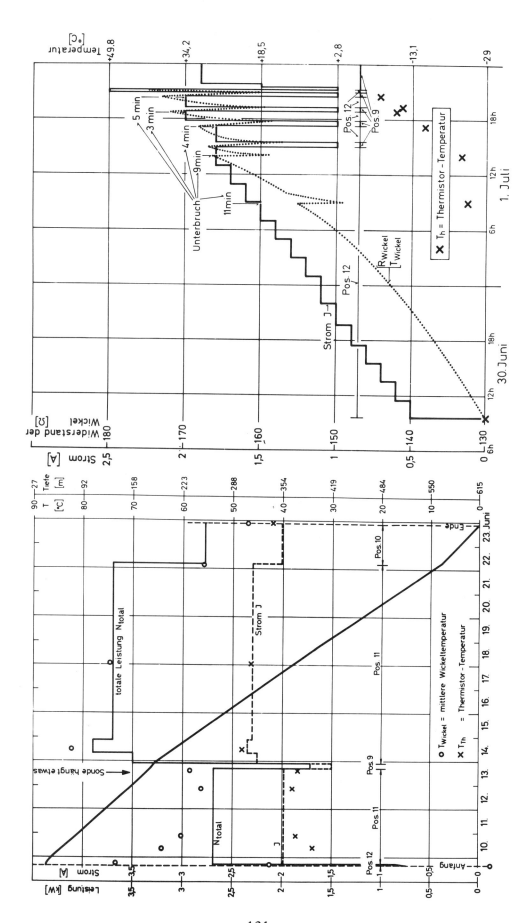

Figure 12. Reheating process of probe II at 615 m depth.

Figure 11. Total power, current and temperatures of coils (mean) and thermistors for probe II between 50 m and 615 m.

131

THE USA CRREL SHALLOW DRILL

John H. Rand

U.S. Army Cold Regions Research and Engineering Laboratory
Hanover, New Hampshire 03755

ABSTRACT

The USA CRREL shallow drill is an electromechanical device designed for continuous coring in firn and ice to a depth of 100 m. The drill bores a 14-cm-diameter hole while obtaining a core 10 cm in diameter at a penetration rate up to 1 m/min in -20°C ice. The cuttings are transported by spiral brush auger flights to a container above the core-storage section. The core and cuttings are removed from the drill after each 1 m run. Additional components include: 100 m of a seven-conductor electromechanical cable, a 6.8-m tower, a hoist which is ski-mounted, and a three-phase 220-V AC gasoline generator. All the equipment has been designed to be transported in a Twin Otter ski-equipped plane and assembled and operated by two men. The total weight of the drill and associated components is 818 kg. The minimum estimated time required to drill 100 m and retrieve core is 15 hours.

Introduction

The USA CRREL shallow drill was designed with the following objectives: To drill as rapidly as possible through firn and ice to a depth of 100 m with a device which could be transported by light aircraft into the field. The design is based upon information and experience obtained over the many years of CRREL's active involvement in ice-core drilling.

The first version of the shallow drill was tested at Station Milcent, Greenland (70°18′ N, 44°33′ W) during the 1973 Greenland Ice Sheet Program (GISP-73) field season. Although the motor and antitorque system proved to be undersized, the principle of having an auger suspended by a cable and driven by a downhole electric motor was valid.

Since the University of Bern, Switzerland, was developing a similar drill the design of the CRREL shallow drill was modified so that it could drill to greater depths (up to 500 m) by increasing the length of the cable. To take advantage of the increased depth capability the overall diameter of the drill was increased so that the existing down-borehole sampling equipment could be used in the hole produced by the shallow drill. This change made it possible to obtain a larger diameter core which is desirable.

During the GISP-74 field season the modified CRREL shallow drill was tested at Station Crête, Greenland (71°07′ N, 37°18′ W). Results from this testing are covered later in this report.

Equipment

The drill and related equipment has been designed for continuous coring in firn and ice to a depth of 100 m. The drill itself is 3.6 m long and weighs 65 kg (Fig. 1). The hole that is bored is 14.2 cm in diameter and the core retrieved has a diameter of 10 cm. Incorporated in the cutting bit are two independent methods of catching the core. The first method is similar to that of the SIPRE hand auger, that is, machined along the inside circumference of the cutting bit is a taper of 1-3/4°. This surface, along with the loose cutting, produces a wedging effect when lifting the drill, thus breaking the core at that location. The second method is similar to the principle used with the thermal drill. This method has two spring-loaded lever arms which cut into the core when the drill is lifted upon the completion of the drilling cycle. The cutting bits and core catcher are connected by four bolts to the inner barrel bottom flange.

The inner barrel consists of a 2.5-m stainless-steel tube, 10.8 cm outer diameter with a 4-mm wall. Two spiral nylon brushes are epoxied on the outside of the tube to form a two-lead auger flight with a pitch of 10 cm. The cuttings produced at the bit are transported by the auger flights to an opening at the top of this section. They are then deflected to the inside of the inner tube. Upon entering the tube they fall to rest on top of the core as it enters from the bottom through the bit. The nylon spirals also act as a bearing surface between the inner and outer barrels.

The outer barrel consists of a steel tube 2.5 m long, 14 cm in outer diameter with a 4-mm wall thickness. The functions of this outer barrel are to provide a continuous surface for chips to move along, to aid in providing a rigid drill required to drill a vertical hole and to help reduce the effective torque that the antitorque system has to react against. The outer barrel is connected to the top of the motor and gear section enclosing the entire bottom length of the drill.

The motor and gear section is composed of a submersible pump motor coupled to a planetary gear adapted from an air motor used for grinding operations.

The specifications of this section are as follows: The motor is a 1.5-hp, three-phase, 220-V, AC, 60-cycle motor which rotates at a speed of 3450 rpm. The gear reducer, which has a ratio of 34:1, reduces the actual rotation of the output shaft to 100 rpm. The gear reducer is attached to the motor with a splined shaft connecting the two units.

The antitorque system consists of three leaf springs, 120° apart. The springs are 3.8 cm wide with a radius of 76 cm. The effective length of engagement with the wall of the hole is 76 cm.

To attach the drill to the cable, a termination is made where a small length of the armor braid is twisted around. The electrical leads continue straight through this termination. A low-melting alloy (Cerro bend) is used to provide a potting compound. This material melts at 70°C, and is poured into a cavity where the twisted armor braid is placed.

During the early stages of the shallow drill's development, the existing thermal drill's base and tower were used. Since then, a base and tower unit has been assembled. This consists of 100 m of cable, a generator, a hoist, and tower.

The electromechanical cable is 0.95 cm in diameter. It consists of two outer layers of steel armor which support the drill, and seven conductors which provide the electrical connection to the motor. The cable is spooled onto an aluminum drum in an orthocyclic winding pattern. The cable end is attached to a slip-ring assembly for continuous transfer of power.

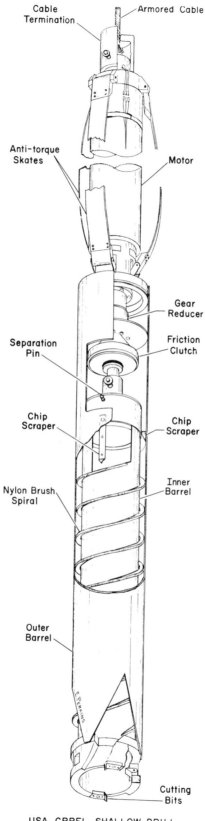

Figure 1. Schematic illustration of CRREL drill.

135

The drum is connected by a chain drive to a gear reducer. On the input side of this gear reducer and in line with a 3-hp motor, 3-phase, 220-V, 60-cycle, AC motor, a clutch is provided for the hoisting operation. The hoisting speed is 60 m/min.

Power for both the electric hoist motor and the drill motor is provided by a 5-kW, 3-phase, gasoline generator.

The tower section is in two parts. The lower section is a telescoping square tubular section. A mechanical screw jack is connected to a variable-speed DC drive motor which raises and lowers the internal section of this tower. This provides a method of controlling the penetration rate of the drill. In addition, it is this system which raises the tower to the vertical position. The top section is an aluminum-extruded tubing which is split lengthwise. This section also serves as a shipping container for the drill.

The entire rectangular aluminum base unit is mounted on three skis. The entire package can be easily disassembled and transported by light aircraft. The total weight of the drill and associated components is 818 kg. The unit can be towed behind a snow vehicle and transported across the snow at a good rate of speed. The entire operation has been designed so that two men can assemble and operate the system.

The drilling cycle starts when lowering the drill down the hole until the bottom is reached. The drill is located just off the bottom as the drill motor is started. The DC drive motor on the tower is started, lowering the tower and drill. Markers on the telescoping tower show when 1 m has been drilled. The drill motor is stopped, the winch slowly raised until core break is felt in the cable. The winch is fully engaged and the drill rises to the surface. Simultaneously the tower is returned to the original position. When the drill reaches the surface the core and cuttings are retrieved and the drill is sent back down again.

Performance

During the GISP-74 field season at Station Crête, several problems developed which made it impossible to continue the testing which was scheduled. Returning to the laboratory, modifications were made to the drill's gear reducer, reducing the final output speed to 100 rpm; several other modifications were made to the winch controls, making the overall system more adaptable for continuous drilling.

In September the drill was shipped back to Greenland for a drilling test at Dye 2. After field changes to the bits, the drill successfully drilled 100 m obtaining an excellent core. The drill was shipped directly from Dye 2 to Antarctica where two holes were drilled (Rand, 1975). The first 100-m hole was drilled in early November at the South Pole under the new geodesic dome. Excellent core was obtained in a record drilling time of 15 hr. Because of the -30°C temperatures the drilling was spread over a 2.5-day period. The second 100-m hole was located at J-9 on the Ross Ice Shelf (Fig. 2).

As mentioned earlier, with the successful drilling attempts of 100 m it is felt that with an increased length in cable and minor alterations to the base, this shallow drill could replace the well known thermal drill in drilling to intermediate depth in firn and ice.

Figure 2. Shallow drill in use during the 1974-75 austral summer field season, Antarctica.

Acknowledgments

I would like to thank Mr. Larry Gould for his efforts during the development stages of the shallow drill, Mr. Robert Bigl and the CRREL machinists for their efforts in the construction of the shallow drill, Mr. Henry Rufli for his assistance, and Mr. B. Lyle Hansen for technical guidance and continual assistance.

REFERENCE

Rand, John H., 1975, 100-meter ice cores from the South Pole and the Ross Ice Shelf: *Antarctic Journal of the United States,* v. 10, no. 4, pp. 150-151.

LIGHTWEIGHT 50-METER CORE DRILL FOR FIRN AND ICE

Heinrich Rufli, Bernhard Stauffer, and Hans Oeschger
Physics Institute
University of Bern
Sidlerstrasse 5
3012 Bern, Switzerland

ABSTRACT

The increasing interest in ice-core analyses has made it apparent that new drills need to be developed for coring in firn and ice.

To close the gap between the range of the SIPRE coring auger and that of the CRREL thermal drill an electromechanical drill has been constructed which will core to 50 m in firn and ice.

The drill is suspended from a cable, and consists of a coring system, a driving system and an antitorque system.

The drill was tested successfully in Greenland in 1974, where it was possible to drill to 50 m in a few hours.

Introduction

With the SIPRE coring auger it is relatively easy to drill to 20 m in firn or 10 m in ice. Drilling deeper is possible but difficult. Holes with depths between 100 and 500 m have usually been made using thermal drills (Ueda and Garfield, 1969).

Attempts have been made to close the gap between the ranges of the two drills by modifying the SIPRE coring auger. The modifications consist of an electrical drive for drilling and electrical and mechanical means for raising and lowering the auger. They allow deeper drilling but the equipment is heavier and the drilling is time-consuming because of the need to handle the numerous extensions. Furthermore, with deeper holes more chips are lost in the hole and the cores are shorter.

Because of these problems the development of a new drill to close the gap seems desirable. The penetration of meltwater into thermally drilled cores in firn makes the cores unacceptable for certain analytical studies. To overcome this problem the new drill should be electromechanically driven.

139

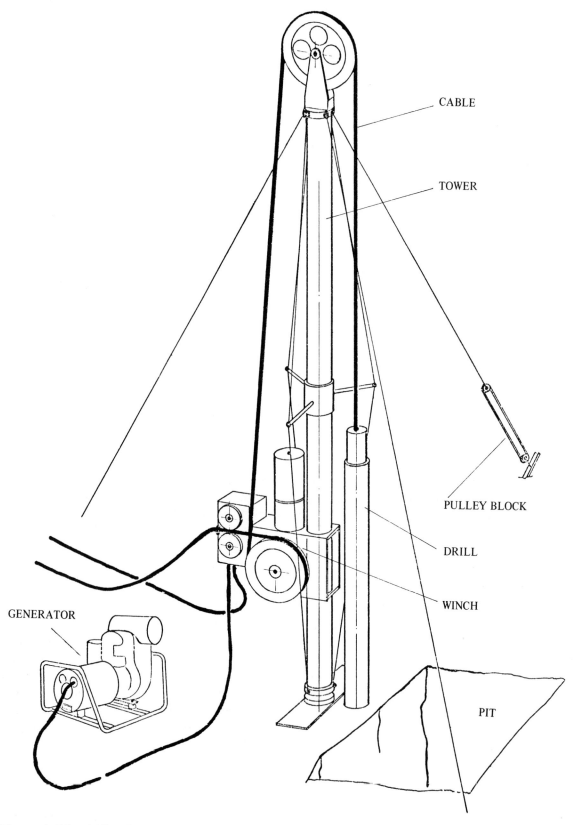

CABLE

TOWER

PULLEY BLOCK

DRILL

WINCH

GENERATOR

PIT

Figure 1. The drill unit.

In the last few years such a drill has been developed at the Physics Institute at the University of Bern, incorporating experience from the SIPRE coring auger and the Icelandic electromechanical drill (Árnason *et al.*, 1974). The prototype was tested and used in the Greenland Ice Sheet Program in 1974 to obtain cores at Summit Station (19 m), Crête (23 and 50 m), and Dye 2 (25 and 45 m).

Description of the Drill Unit

The drill unit consists of four parts (Fig. 1): drill with cable, tower, winch, and generator.

No single part is heavier than 50 kg, so that the equipment can be transported in any type of helicopter or on a small sledge. The total weight of the complete unit is 150 kg.

The Drill (Fig. 2)

A core barrel, similar to the SIPRE coring auger but about 2 m long with an auger flight over the total length, rotates inside an outer (jacket) tube. Four holes in the upper half of the core barrel provide inlets for the ice chips which are transported up by the auger flights between the core barrel and the outer tube and get packed behind the core. After each run (70 to 90 cm long) the core barrel is separated from the rest of the drilling unit and the core and the chips are removed.

The drill has the following parts, each of which is described in detail: cable and cable connection, antitorque system, driving section, and coring section.

Cable and cable connection:

A reinforced electrical rubber-jacketed cable with a load capacity of 300 kg is used. It consists of seven electrical conductors (cross section 1.5 mm^2) and is reinforced with three hemp ropes. The cable is covered with a special rubber which remains flexible down to -35° C. The cable diameter is 17 mm and the weight is 39 kg per 100 m with a bending radius of 120 mm. The cable is fastened in a cable termination with epoxy (cast) resin.

The antitorque system (Fig 3):

This system prevents the rotation of all parts of the drill except the core barrel and its driving mechanisms. This system has to work in loose firn as well as in hard ice. To accomplish this two devices are used. In the first device 4 knives (skates) which are retracted during the lowering or raising of the drill are pressed radially against the wall of the borehole as soon as the drill reaches the bottom of the borehole. This device is effective in hard firn or ice but the knives (skates) are not long enough for soft firn. The second device consists of 2 plate springs which press against the wall of the borehole at all times. They provide the additional restraint required in the soft firn.

The driving section:

The drill is driven by an air-cooled electric commutator motor which uses 500 W, 200 V AC

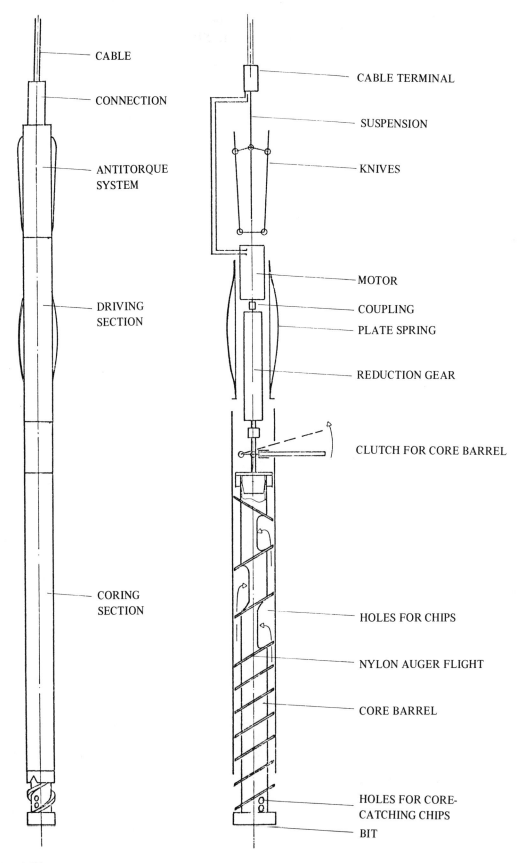

CABLE

CONNECTION

ANTITORQUE
SYSTEM

DRIVING
SECTION

CORING
SECTION

CABLE TERMINAL

SUSPENSION

KNIVES

MOTOR

COUPLING

PLATE SPRING

REDUCTION GEAR

CLUTCH FOR CORE BARREL

HOLES FOR CHIPS

NYLON AUGER FLIGHT

CORE BARREL

HOLES FOR CORE-
CATCHING CHIPS

BIT

Figure 2. The drill.

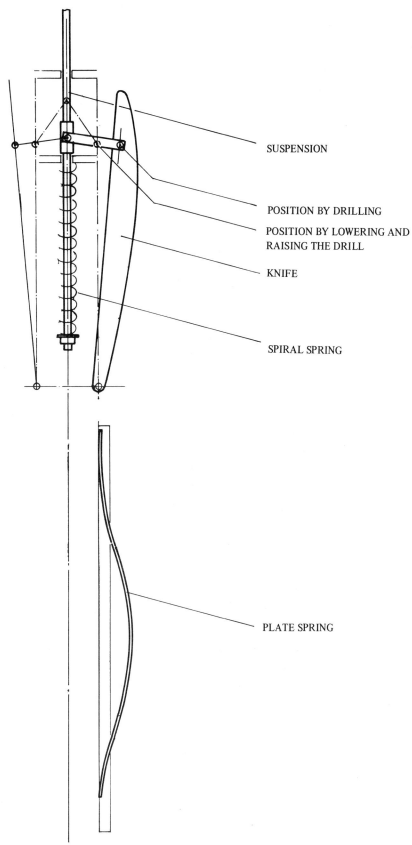

SUSPENSION

POSITION BY DRILLING

POSITION BY LOWERING AND
RAISING THE DRILL

KNIFE

SPIRAL SPRING

PLATE SPRING

Figure 3. The antitorque system.

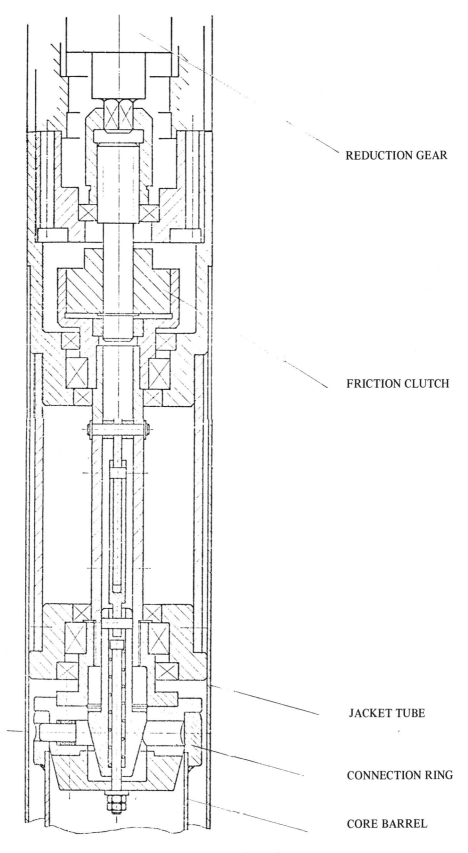

REDUCTION GEAR

FRICTION CLUTCH

JACKET TUBE

CONNECTION RING

CORE BARREL

Figure 4. The clutch between driving section and core barrel.

with 50 or 60 c/s. The mechanical power at 23,000 rpm is 320 W. The high speed of the motor is reduced by planetary gears to 90 rpm. Between gear and core barrel a friction clutch (Fig. 4) prevents overloading of the gear reducer and motor.

The coring section (Figs. 5 and 6):

The coring section consists of a core barrel (Fig. 5) which fits inside the non-rotating jacket tube (Fig. 6). The core barrel, a 2-m-long steel tube, has an inside diameter of 80 mm. The lower half of the core barrel has a double-auger flight which raises the chips and serves as a bearing for the core barrel in the jacket tube. The auger flights are made of nylon attached to the core barrel with screws. Only one flight goes over the upper half of the core barrel which has four holes to admit chips to the space above the core. A steel ring is welded to the top of the core barrel to connect it to the driving section. The drill bit is attached to the core barrel with screws. It is fitted with two replaceable knives similar to those on the SIPRE coring auger but adjustable in diameter. The inner diameter, 75 mm, is also the core diameter; the outer diameter, 114.5 mm, cuts a hole whose diameter is 115 mm. Four 2-cm-diameter holes above the bit allow chips to enter the tapered space between the core and the bit where they provide a core-catching action like that used on the SIPRE coring auger. A movable styrofoam disc was placed in the middle of the core barrel after each run to separate the core and the chips.

The jacket tube (Fig. 6) covers the entire length of the core barrel except the lower 15 cm. Three ribs along the inner side provide the frictional resistance between the spinning chips and the non-rotating jacket tube that is required for the auger flights to elevate the chips to the inlet at the top of the core barrel. The bottom of the jacket tube has an inlet ring with three triangular notches. Its outer diameter, 113 mm, is just 2 mm less than the diameter of the borehole.

The Tower (Fig. 1)

The tower consists of two aluminum tubes, each 2.3 m long, 130 mm O.D. and 124 mm I.D. On the top is the sheave for the cable. The bottom has a ball joint (Fig. 7) which provides a movable platform for the tower. The tower is 5 m high and is held erect by three guy-wire cables. All three cables have a pulley block with a nylon rope at the ground anchor end. Using the pulley blocks and the ball joint of the tower, it is easy to bring the tower into a vertical position, even in steep terrain.

The Winch (Fig. 8)

The winch is fixed to the tower. The cable on which the drill hangs goes over the sheave on the top of the tower, then down through the winch and is spread out on the surface because there is no cable drum. The winch has no reel, but works similar to a capstan. An electric motor, single-phase 220 V/550 W, drives a double-gear system (variable reduction gear and then a worm gear). The first gear is variable between 0 and 1100 rpm, and the maximum speed after the second gear is 38 rpm. The winch also includes the control panel. An ammeter guards against an overload of the drill motor and the winch motor.

The Generator

A two-stroke gasoline engine drives a single-phase 2-kW generator. The unit is fixed on a

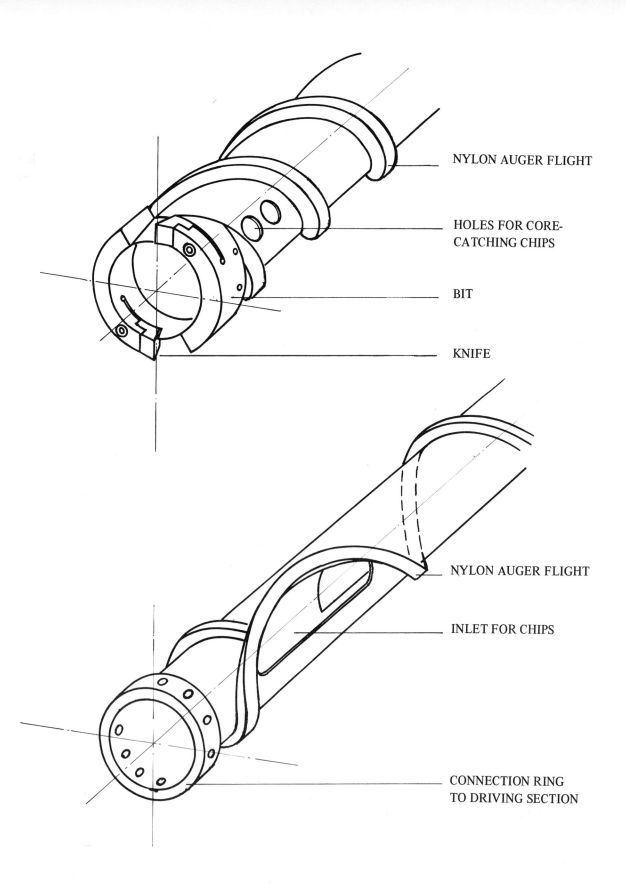

NYLON AUGER FLIGHT

HOLES FOR CORE-CATCHING CHIPS

BIT

KNIFE

NYLON AUGER FLIGHT

INLET FOR CHIPS

CONNECTION RING TO DRIVING SECTION

Figure 5. The core barrel.

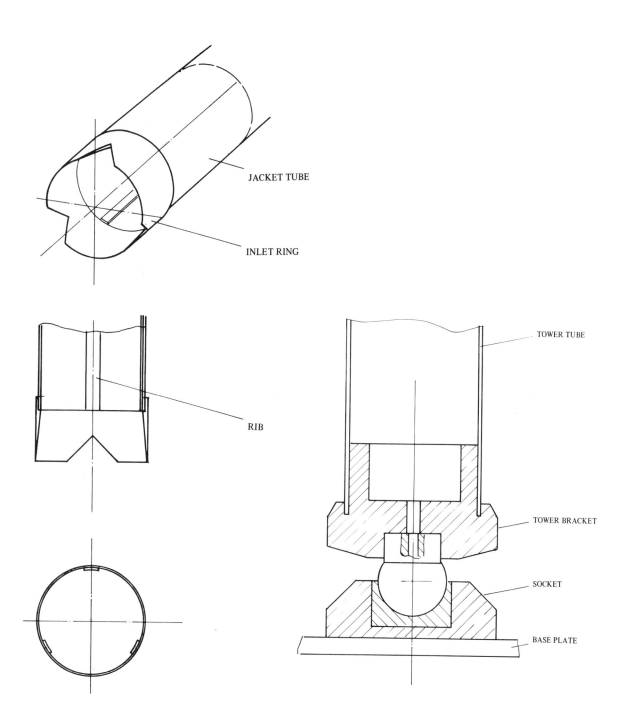

JACKET TUBE

INLET RING

RIB

TOWER TUBE

TOWER BRACKET

SOCKET

BASE PLATE

Figure 6. The jacket tube.   **Figure 7**. The ball joint.

147

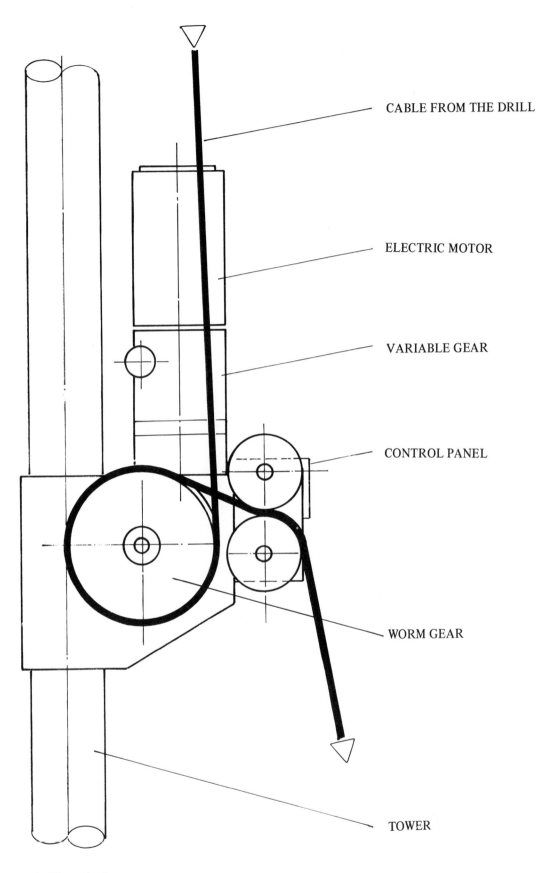

CABLE FROM THE DRILL

ELECTRIC MOTOR

VARIABLE GEAR

CONTROL PANEL

WORM GEAR

TOWER

Figure 8. The winch.

frame and its weight is 38 kg. The fuel consumption is about 5 gallons of gasoline for a 50-m borehole.

Drilling Preparation

For transportation the complete drill unit is packed in three boxes with a total volume of 0.85 m^3. It takes two men about two hours to assemble the drill and to get it ready for drilling. One man digs a pit about 1.5 m on a side and 1.5 m deep to provide room to pull the core barrel down from the drill, while the other man assembles the tower, the winch, and the cable. To raise the tower, one man holds the ball joint firmly on the surface and the other raises the tower like a long ladder. While one man holds the tower vertical, the other fixes the anchors in the snow and adjusts the pulley blocks. When the drill is connected to the cable it serves as a plumb to bring the tower into its final perpendicular position.

The generator is normally located about 10 to 15 m from the drill tower.

Drilling Operations

At the beginning of drilling, it is very important to maintain a vertical hole. Therefore for the first 3 to 5 cores the drill feed is kept very slow. After the first 5 m of drilling, it is helpful to dry the drill on the outside and let it sit in the hole for awhile, allowing it to cool to the temperature of the firn.

The complete run involves the following steps:

1. Lower the drill down the hole at the full speed of the winch.

2. Begin decelerating a few meters above the bottom of the hole and touch the bottom very slowly.

3. Give slack on the cable corresponding to the length of the new core to be drilled.

4. Stop the winch and start the motor of the drill.

5. Hold the cable over the hole by hand. This gives the best feeling for the feed of the drill.

6. After the desired length is drilled, switch off the drill motor.

7. Start the winch to raise the drill with very low speed.

8. Pull on the cable by hand to break the core.

9. Increase the speed of the winch and raise the drill.

10. Disconnect the core barrel.

11. Push the core out.

12. Replace the disc which separates the core barrel into a core and chips section, and clean the barrel with a brush.

13. Re-connect the barrel.

14. Check the drill function with the motor.

A run at 40 m depth takes about 5 to 6 minutes.

Drilling Experience and Tests

Tests were made in 1973 at Dye 2, Greenland, in February and March 1974 in the Swiss Alps on Jungfraujoch and Plaine Morte, and in 1974 in Greenland at Summit, Crête, and Dye 2. In 1973 we had a complete drill unit except for the electric winch. After some trouble with the coring section, we changed it and used the SIPRE auger together with the driving section. This allowed us to get some experience with the driving system and the drill tower. With this combination we drilled to 24 m. To move the drill up and down we used a hand-driven winch.

In the 1973-74 winter we built a new coring section and an electric winch. We tested the new core barrel on Jungfraujoch in February 1974 by hand driving, just to find out how the new principle was working.

Alpine Glacier

In March 1974 the new core barrel, together with the driving unit, was tested on Plaine Morte, a glacier in the Alps 2700 m above sea level, and got the first proper experience in snow and glacier ice with the new drill. We still used the hand-driven winch. Several short holes were drilled through the 3-m snow layer on the ice and down to 5 m into the ice. One of the most important considerations was to determine the best diameter for the bit. For this purpose we had different sets of cutters with us. We also modified the inlet ring of the jacket tube and made some torque measurements to find out the power required to drill in ice. The drilling speed we got in snow was 2 cm/sec, and in glacier ice 1 cm/sec. With a core barrel length of 210 cm, we got 100- to 120-cm cores in snow, and 60- to 70-cm cores in glacier ice. The inlet ring of the jacket tube was modified by cutting notches (Fig. 6) so that the chips would enter the jacket tube while drilling in ice. Without the notches the chips were merely rotating with the core barrel on its open end and would get packed in front of the jacket tube, giving us trouble in raising the drill. The free part of the core barrel was 25 cm at that time, but we later made it shorter. The rest of the new coring section did not have to be changed.

Greenland Firn (1974)

Summit (Fig. 9):

Except for some trouble with the inlet ring, the drill worked well in principle. At a depth of 19 m the coupling between the motor and the reduction gear broke and could not be repaired in the field.

Figure 9. Drill unit assembled at Summit Station, Greenland.

Crête:

After a complete overhaul of the drill we tried to reach 50 m depth at Crête Station.

Antitorque system:

At a depth of 23 m the entire drill began to rotate. With the increasing density of ice the friction increased between the core and the bit because the cutters were exactly flush with the inlet diameter of the bit so as to guarantee good core catching. To reduce the friction we moved the cutters about 0.1 mm each to the inside and got better results. We also fixed a helical spring behind the plate springs of the antitorque system (Fig. 3) to give them more pressure against the wall of the borehole.

Even after these changes we still could not penetrate beyond 23 m depth. We then started a new hole and succeeded to reach the desired depth of 50 m.

151

Cooling the drill before drilling:

On the surface the drill is normally warmer than the firn in the borehole because it is warmed by the air and the solar radiation. The warm drill gets wet in the borehole and becomes covered with ice, and the chips freeze to the core barrel. To avoid this it is best when beginning a new hole to drill only the length of the drill. Then dry the drill including the core barrel, lower it into the hole and let it cool well below the melting point of ice before resuming drilling. When there is a lull in the drilling always leave the drill in the hole.

Core catching:

When drilling the second hole we had trouble at a depth of 30 m with core catching. The cutters made the core surface too smooth and the chips too fine (almost powder). A change in the angle on the inside of the cutters produced grooves in the surface of the cores, and we could catch them again, even at a depth of 50 m where the firn has a high density.

Winch:

The winch was not as strong as it was planned to be. The maximum speed of 25 m/min was never reached in raising the drill.

Drilling techniques:

One of the most important results was the drilling technique outlined above (Drilling Operations). The total drilling time at Crête for the 50.50-m hole was about 10 hours.

Dye 2:

Because Dye 2 is in the percolation zone, we had different firn conditions from those at Crête.

We drilled to 25 m in 3 hours, when the drill stuck downhole and we were unable to raise it with the winch. We tried to raise it with glycol, but after 2 days without results we made a final effort. A 200-liter drum of glycol was heated to 80° C with an electrical heater and the hot glycol was put into the hole. The drill was freed very easily, and we could see the reason why it stuck. Two pieces of wood that were placed on the inside of the plate springs at Crête absorbed meltwater and swelled, so the springs could not move in.

After a complete overhaul of the drill, it was seen that nothing was damaged by the glycol, but we had to start a new hole. We drilled 45 m in 7.5 hours without any trouble. No cores were lost, and the average length of the cores was 80 cm.

Conclusions

The drill unit described seems to work well for shallow-hole drilling in firn and ice down to 50 or 70 m, but some modifications to the drill and the winch are necessary. To drill in temperate glaciers the drill needs to have a submergible driving section. The antitorque system will require longer skates instead of the present plate springs.

Acknowledgments

This work was supported by grants from the Swiss National Science Foundation and the U.S. National Science Foundation. We thank Mr. B. Lyle Hansen and Mr. John Rand from CRREL for their advice concerning the design; Mr. P. Domke, his apprentices, and Mr. W. Bernhard for constructing the prototype; Dr. John Röthlisberger for his help on Plaine Morte; and Dr. Sigfus Johnsen for his help in GISP-74.

REFERENCES

Árnason, B., H. Björnsson, and P. Theodórsson, 1974, Mechanical drill for deep coring in temperate ice: *Journal of Glaciology,* v. 13, no. 67, pp. 133-139.

Ueda, H.T. and D.E. Garfield, 1969, The USA CRREL drill for thermal coring in ice: *Journal of Glaciology,* v. 8, no. 53, pp. 311-314.

Note added in proof: A new and modified drill was built in 1975 and used to drill four holes for the Greenland Ice Sheet Program-1975: a 94-m hole at Dye 3, an 80-m hole and a 30-m hole at South Dome (about 190 km south of Dye 3), and a 60-m hole at Hans Tausen Ice Cap. The new drill is designed for drilling to 100-m depth and the complete unit weighs about 350 kg, including tool box. The drill and tower are packed in one box (2.4 m long, 0.28 m^3, 90 kg), and the remaining components (less tools) in four boxes (each 0.14 m^3, 60 kg). Two winch systems were made for the drill: one is a capstan winch with 120 m of rubber cable, and one is a conventional winch system with a drum and a steel cable with three conductors. Both winch systems use the same motor and gear box. The winch speed is about 40 m/min and the motor has a thyristor driving unit for the speed variation.

DEEP CORE DRILLING

BY JAPANESE ANTARCTIC RESEARCH EXPEDITIONS

Yosio Suzuki
Institute of Low Temperature Science
Hokkaido University
Sapporo 060, Japan

ABSTRACT

Deep drilling by Japanese Antarctic Research Expeditions in 1971 during JARE XII began at Mizuho Camp (70°42.1' S, 44°17.5' E) with a 400-m winch, a 2.4-kW thermal drill and a 100-W electrodrill. In 1972, JARE XIII reached 147.5 m with a new thermal drill. Plans were made in 1973 to reach 800 m by February 1975, so JARE XV installed a new 800-m winch at Mizuho in May 1974. Drilling will start in October 1974 with a 3-kW thermal drill. In January 1975, two people from JARE XVI will join the operation with another 3-kW thermal drill. A thermal drill similar to JARE XV's and a 400-W electrodrill were tested in November 1973 at Ice Island T-3 by a party from Nagoya University, who successfully obtained 30 m of 250-mm-diameter cores using those drills and a large 6-kW thermal drill. JARE programs have thus made two winches, five thermal drills and two electrodrills, and one thermal drill is now in preparation.

Introduction

Despite active participation of glaciologists in JARE, no deep drilling project had been proposed by Japan until 1969. One of the reasons was that JARE had no base on the continent or on an ice shelf. Also, until JARE IX, Japan's main concern had been directed toward a traverse between Syowa Station and the South Pole by JARE IX in 1968-69.

Prior to JARE IX's departure from Japan in 1967, a long-range glaciological research plan covering JARE X, XI, XIV and XV had been approved, in which extensive oversnow traverses were scheduled over an area from 35° E to 52° E, requiring a permanent depot at about 71° S, 45° E.

Soon after JARE X departed Japan, several glaciologists, including the author, proposed to use the depot as a drilling site by JARE XII, XIII and XVI. In May 1969 the drilling project was approved and incorporated with the traverse project to form the Mizuho Plateau-West Enderby Land Project covering JARE X through XVI.

The transportation capability of JARE limited the total weight of drilling equipment to 1000 kg. Based on available information (Patenaude *et al.*, 1959; Shreve and Kamb, 1964; Ueda and Garfield, 1968, 1969a, 1969b), plans were made to reach 400 m with a thermal drill by 1972, and JARE XVI was to reach 1000 m with an electrodrill, yet to be developed. With the courtesy of CRREL, a full set of blueprints of a winch and a drill, CRREL Mk II, was obtained, and feasibility studies of making them in Japan were started in August 1969.

A budget of about $10,000 was allocated to the drilling project in fiscal 1970. A 400-m cable made in accordance with CRREL specifications and a 1.5-kW winch were ordered. Because of difficulties in obtaining thick aluminum and plastic pipes, the drill design was greatly changed and a small 2.4-kW drill, JARE 140, was made. As a trial, a 100-W electrodrill was also made. Meanwhile, JARE XI opened a depot, Mizuho Camp, at 70°42.1′ S, 44°17.5′ E in July 1970. The ice thickness there was reported as 2095 m.

JARE XII personnel arrived at Mizuho on June 28, 1971, and stayed for two weeks to install a 12-kVA generator and to construct living quarters. They returned to Mizuho at the end of September with the drilling equipment. After installing the winch in a 4-m pit, they began drilling on October 16 with the electrodrill, which stuck at 38.8 m on November 1. Recovery efforts failed on November 6, when the cable slipped out from the clamp of the drill. Drilling of a new hole began the next day with the thermal drill, which was lost at 71 m on November 17.

It had been hoped to develop a practical electrodrill for JARE XIII, but because no motor with an appropriate gear reducer was found, a thermal drill was adopted again. A new drill, JARE 140 Mk II, designed for easy disassembly, was used by JARE XIII personnel, who started drilling in July 1972. Despite many problems, they reached 104.5 m on September 14, when the drill stuck. By pouring 60 liters of antifreeze into the hole they recovered the drill, but only with severe damage to the pump. A new pump was sent from Syowa and they restarted drilling on November 6, reaching 147.5 m on November 14, where the drill again stuck and was abandoned.

Because low ambient temperatures in the drilling operations of JARE XII and XIII caused many problems, it was decided that the operation of JARE XVI should be done in January-February 1975, with personnel being flown to Mizuho from the ship. Equipment had been transported by JARE XV in the previous year. Since it would only be possible to work a maximum of 40 days, the target depth was lowered from 1000 m to 800 m. Preparation was begun of a winch with 800 m cable, two 3-kW thermal drills (JARE 160), and an electrodrill.

Failures in JARE XII and XIII to reach the target depths were partly attributed to the lack of field tests of the drills. Fortunately, Nagoya University was planning to take large ice cores from T-3 in October, so it was decided to test the new drills there.

The Nagoya party arrived at Barrow, Alaska, in September 1973 with one of the JARE 160 drills, the electrodrill and a large thermal drill, Type 300. Poor weather prevented them from getting to T-3 until the end of October, but by the end of November, they succeeded in getting 30 m of 250-mm-diameter cores and 31 m of 132-mm-diameter cores with the thermal drills.

Based on the interim report from T-3, the other JARE 160 drill was modified into the JARE 160A, which, together with the 800-m winch, was sent on JARE XV. The winch was installed at Mizuho in May 1974. The test drilling is due to start in October. In January 1975, two people from JARE XVI will join the operation with a new thermal drill, JARE 160B, which is now in preparation.

The Japanese ice-drilling activities to date are summarized in Table 1.

Table 1

Japanese Ice-Drilling Activities

Party	Year	Equipment	Performance
JARE XII	1971	12-kVA 200-V 3-phase generator 1.5-kW winch with 400 m cable 2.4-kW thermal drill (JARE 140) 100-W electrodrill	38.8 m by electrodrill 71 m by thermal drill
JARE XIII	1972	2-kW thermal drill (JARE 140 Mk II)	147.5 m
Nagoya University	1973	6-kVA 200-V 1-phase generator 50 m electric cable Gasoline-powered 100-m winch 3-kW thermal drill (JARE 160) 400-W electrodrill 6-kW thermal drill (Type 300)	30 m of 250-mm-diameter cores 31 m of 132-mm-diameter cores
JARE XV	1974-75	12-kVA 200-V 3-phase generator Transformer and rectifier to supply DC 0-230 V 30A 1.5-kW winch with 800 m cable 3-kW thermal drill (JARE 160A)	
JARE XVI	1975	3-kW thermal drill (JARE 160B)	

Description of the Components

(A) *Power Sources:* A 5-kW 200-V single-phase AC generator was considered as a power source in the early stage of planning, but it was later decided to install a 12-kVA 200-V three-phase AC generator at Mizuho for general use. While a three-phase generator allows the use of an easily-obtained three-phase induction motor for a winch, it presents difficulties in powering a drill through a cable which is intended to transmit single-phase current (see below). Namely, to draw a large power load from one phase of the generator causes a severe imbalance, thus requiring dummy loads on other phases and increasing fuel consumption considerably. As a remedy, JARE XV brought a rectifier which converts three-phase AC into DC current, possibly solving the problem. At T-3, a 6-kW 200-V single-phase AC generator was used.

(B) *Cables:* Specifications of the 400-m cable made in 1970 and the 800-m cable made in

1973 are shown in Table 2. Both are CRREL-type cables, having seven control conductors and one power conductor, the latter together with armors being intended to transmit DC or single-phase AC current. Extensive laboratory testing of the 1970 cable showed it to be weak in radial impact. When a 20-kg weight was dropped on the cable from a height of 0.5 m, the insulation broke down on some of the control conductors. Such a radial impact can occur during drilling when the cable slips off a sheave. To prevent this, the groove of the sheave must be deep enough. Because the cable undergoes repeated bending during drilling, repeated bending tests were carried out in low ambient temperatures (Fig. 1).

Table 2

Specifications of Cables

	JARE XII (400 m)	JARE XV (800 m)
Plain Copper: Number/dia.	7/0.23 mm	7/0.23 mm
Thickness of Nylon Insulation	0.23 mm	0.23 mm
Diameter of One Conductor	1.15 mm	1.15 mm
Diameter of Seven Conductors	3.45 mm	3.45 mm
Thickness of Mylar Tape Layer	0.15 mm	0.15 mm
Plain Copper: Number/dia.	12/1.18 mm	15/0.9 mm
Diameter up to Power Conductor	6.1 mm	5.55 mm
Thickness of Polyethylene	0.85 mm	0.8 mm
Thickness of Braid	0.3 mm (Nylon)	0.3 mm (Polyester)
Galvanized Steel: Number/dia.	12/1.0 mm	14/0.8 mm
Tinned Hard Copper: Number/dia.	12/0.99 mm	14/0.8 mm
Diameter up the First Armor	10.4 mm	9.35 mm
Galvanized Steel: Number/dia.	27/1.20 mm	25/1.20 mm
Diameter of Cable	12.8 mm	11.8 mm
Resistance: Control Conductor	< 65 Ω/km	< 70 Ω/km
Power Conductor	< 1.4 Ω/km	< 2.0 Ω/km
Armor	< 1.9 Ω/km	< 2.5 Ω/km
Tensile Strength	> 4000 kg	> 3000 kg
Weight	650 kg/km	450 kg/km

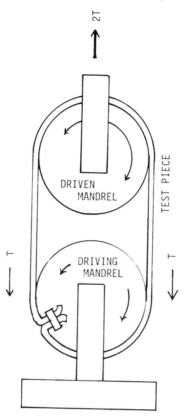

Figure 1. Schematic diagram of the bending test of the cable.

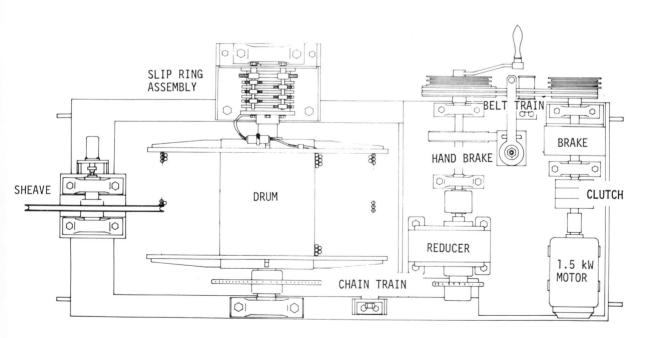

Figure 2. Plan of the 1973 winch.

The test piece of the cable, half-way wound on a mandrel of 320-mm diameter, is forced to move back and forth while tension is applied to the cable. With a 1000-kg tensile force, -25° C ambient temperature and 33 cycles per minute reciprocal movement, the cable showed no serious damage at 2000 movements while at 3000 movements some copper wire in the inner armor broke and pierced the polyethylene layer, causing a short circuit to the power conductor. The tensile strength of the cable was over 4800 kg, enough to hold the weight of the cable itself, 260 kg, plus that of the drill, about 50 kg, and the breaking strength of the ice core, estimated at several hundred kilograms.

In order to ensure the minimum winding speed of 20 m/min of the winch, the weight of the 1973 cable was considerably reduced at the sacrifice of the tensile strength and the conductivity of the power conductor. The designed value of 3000 kg of the tensile strength seems enough for breaking ice cores, and the lower conductivity can be overcome by use of high voltage, if necessary.

(C) *Winches:* The 1970 drum was similar to the CRREL model with 7 elements of plane ring slip rings, which was replaced by cylindrical ones in the 1973 drum, because the latter is easy to make and to maintain. The 1970 drum with 400 m of cable weighed about 450 kg, while the 1973 drum with 800 m of cable weighed about 550 kg.

The power trains are similar for both winches (Fig. 2). A 1.5-kW three-phase motor was used, which might be better replaced by a DC motor, if the rectifier mentioned above works well. The electromagnetic clutch and brake were adopted in anticipation of automatic control by such signals from the drill as cable tension, water level in the tank, etc., but no automatic control was actually used.

The frame and mast were made of steel instead of aluminum alloy. The height of the mast was shortened to 3.5 m for both winches.

(D) *Thermal Drills Other than Type 300:* Specifications are given in Table 3, including the Type 300 drill and the CRREL drill (Ueda and Garfield, 1969a, 1969b). The working principle is the same as the CRREL drill. Each consists of five blocks: spring suspension block, vacuum pump block, water tank block, core barrel block and main heater.

(D-1) *Spring Suspension Block* (Fig. 4): The cable was fixed to the inner cylinder with a screw clamp instead of with plastic (epoxy) cement or low-temperature alloy (white metal). The screw clamp was adopted for easy reattachment of the cable in case it breaks. The actual clamping force was not measured. Specifications of the suspension spring and the load-indicating system were the same for all drills. Three ranges of the load—small, normal and large—were indicated on the surface control panel by signals from a microswitch assembly at the suspension block. The load in normal range was between 14 and 18 kg.

(D-2) *Vacuum Pump Block:* A diaphragm-type 20-W vacuum pump, IWAKI AP 220, capable of producing a vacuum of -450 mm Hg, was used throughout, with modifications to fit in a 130-mm-diameter (for the 140) or 150-mm-diameter (for the 160) cylinder. In low ambient temperatures, the diaphragm often stiffened, making the pump difficult to start. A heater (Fig. 3) was added to warm the pump for the 160A and 160B.

The pump housing of the JARE 140 was a stainless-steel cylinder with a cover on top and a female screw thread at the bottom. The suspension was bolted to the top cover, while

Table 3

Specifications of Thermal Drills

Type	Length, mm	Weight, kg	Core Capacity Dia./Length, mm	Heater Ring O.D./I.D./Height, mm	Heater Elements	Estimated Melting Area, cm^2	Power Per Unit Area, W/cm^2
140	2500	30	103/1000	142/105/75	100 V 1.2 kW x 2	80	30
140 Mk II	3050	40	105/1200	142/108/75	100 V 1.0 kW x 2	75	27
160 & 160A	3420	50	132/1500	168/134/70	200 V 1.5 kW x 2	90	33
160B	4000	60	132/2000	168/134/65	100 V 1.5 kW x 2	90	33
300	2080	140	250/1500	285/252/100	200 V 2.0 kW x 3	180	33
CRREL Mk II	4600	80	122/1500	162/124/51	215 V 625 W x 18 Used at 115 V ca. 3.2 kW	90	36

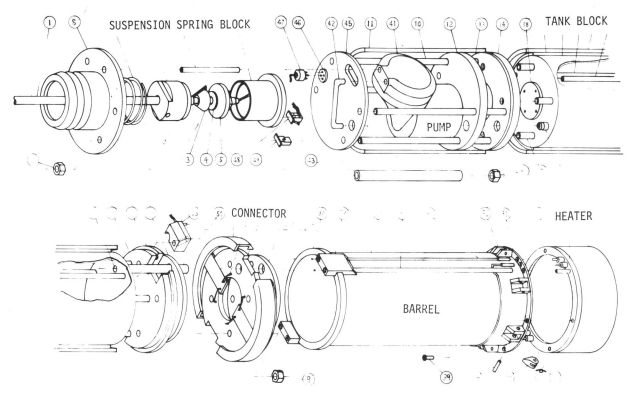

Figure 3. The JARE 160A thermal drill.

a male coupler, glued to an FRP casing of the tank and barrel block, fitted in the female screw. The housing of every other drill was a "cage" made of four long bolts planted on a basal disk with an FRP casing which was not a tensile structural member. The suspension was fixed to the bolts and the disk was fixed to the tank (see below).

(D-3) *Water Tank Block* (Fig. 5): Every tank was made of a 2.1-mm-thick stainless-steel pipe, of either 114.3 mm O.D. (the 140) or 139.8 mm O.D. (the 160). A simple water gauge using a buoy was mounted in every tank. The 160B tank will have an electromagnetic valve to recover its inner pressure quickly so as to make water in the suction tubes flow down rapidly (see next paragraph). Every tank was covered with foam plastic, and inserted in 4.5-mm-thick FRP pipe of either 139 mm O.D. (the 140) or 159 mm O.D. (the 160), except for the 160A tank, which was covered with thin sheet steel lined with a rubber sheet heater. The heater is intended to be used if the drill freezes in the hole. Every tank except the JARE 140 was a structural member with four bolts fixed on each end panel. The upper four bolts fastened the base of the pump housing on the tank, while the lower four fixed a connector (Fig. 6) for the barrel block on the tank. The connector of the 140 Mk II was a simple disk with six holes. That of the 160 was a sophisticated device with two semi-circular arms, which, when closed, gripped the flange of the barrel so as to fix the barrel to the tank.

(D-4) *Core Barrel Block* (Fig. 7): The barrel of the JARE 140 was an extension of the FRP tank casing. A core-cutting ring was glued to its lower end. A special feature of the ring is its catcher-releaser, whose turning can instantly release all catchers from the core so that the core is easily taken out downward.

162

The structural member of the JARE 140 Mk II barrel block was six 6-mm-diameter shafts with a male screw threaded on both ends. The shafts were fixed on a core-cutting ring similar to the 140. An inner case was inserted between the shafts. Water piping and electrical wiring were fastened to the inner case, as were the shafts. A 139-mm-O.D. FRP pipe covered them for protection. The shafts were fastened with nuts on the holes of the disk of the tank block. The above construction made it easy to disassemble the barrel block for inspection of water piping and electrical wiring.

The barrel of every 160 series drill (Fig. 3) was a 2.1-mm-thick stainless-steel pipe, 139.8 mm O.D., with core-catchers welded at the lower end and a flange at the upper end. The flange was for the connection of the barrel to the tank block (see previous paragraph). The barrel had no catcher-releaser. Core was taken out by detaching the barrel from the tank block and turning it upside down.

Water tubes for all the drills were stainless steel (6 mm or 8 mm O.D.). A silicone-rubber lead heater of 2-mm diameter was used for warming the tubes, except for those of the 160B, which were hoped to have no freezing problem because of the pressure release valve of the tank.

(D-5) *Main Heaters:* To avoid the fine machining required by the CRREL design, where cartridge heaters were inserted in an aluminum ring, we adopted molded heaters throughout. The heating elements were made of stainless-steel sheath heater of 8-9 mm diameter.

(E) *Type 300 Thermal Drill:* The drill consisted of a core barrel with a heater and a side pilot pipe with a core cutter. Meltwater was allowed to flow in a pilot hole previously drilled by the JARE 160, and was occasionally taken out with a bucket. A possible version is to use the lower part of the pilot pipe as the water tank.

(F) *Electrodrills:* The 1970 drill consisted of four blocks: a cable suspension, an anti-torque device, a power unit, and a barrel. The suspension was common with the JARE 140 thermal drill. The anti torque device was of the pantagraph-type with four pairs of arms expanded outward by four adjustable springs. The power unit was a 100-W 100-V single-phase 4-P induction motor with a 15:1 gear reducer. Thus, at 50 Hz the drive shaft rotated at 100 rpm. The barrel was 1.5 m long and was made of 114.3 mm O.D. steel pipe. At the lower end of the pipe was welded a cutter shoe of 150 mm O.D. and 105 mm I.D. on which three cutters were fastened with hexagonal bolts. Triple spiral fins were welded all over the barrel. The pitch was a uniform 150 mm. Hence, two adjacent fins were 50 mm apart vertically. The upper half of the barrel was the reservoir for ice chips. In actual operation, the drill revealed many defects, the most serious one being the ineffectiveness of the spiral fins to move ice chips upward. Often only 20 or 30 cm of drilling caused ice chips to cling between fins and the wall of the hole, overloading the motor. The loss of the drill at 38.8 m was due directly to this defect. Another defect was the difficulty of adjusting the anti torque spring. This might be overcome by adding "skates" to the arms. The power was felt to be slightly insufficient.

The 1973 drill was primarily made for the test at T-3. The drill had no complicated suspension devices, but rather a simple hook on its top. A 200-V 400-W single-phase 2-P motor (3000 rpm by 50 Hz) was mounted on the upper base while a 39:1 gear reducer was fixed at the lower base of an anti torque device. The motor and the reducer were coupled by a spline mechanism to accommodate the change of the height of the device, which was of the pantagraph-type with three pairs of arms. "Skates" were not added, though later they were felt necessary. The weight of the motor was considered enough to expand the arms, but in actual test it was not. The barrel

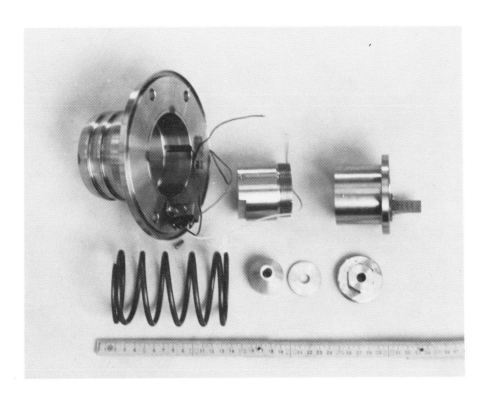

Figure 4. Suspension block.

Figure 5. Upper end of tank block and frame of pump housing. Two limit switches are for monitoring vacuum and water level.

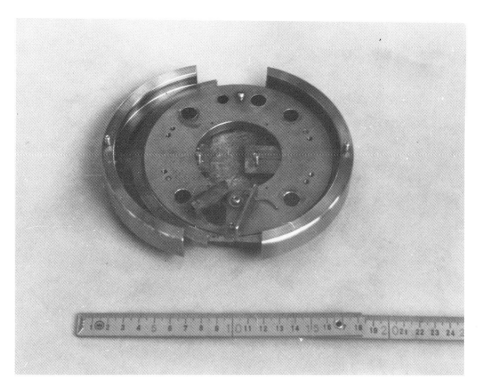

Figure 6. Lower side of connector.

Figure 7. Lower part of core barrel and heater.

165

was made of stainless-steel pipe of 139.8 mm O.D. and 134 mm I.D. The cutting shoe was of 165 mm O.D. and 131 mm I.D. Two cutters were fastened on the shoe, each by one hexagonal bolt. A special feature of the shoe assembly was that it had two claws for core-breaking. When the drill rotated in reverse, the claws caught and struck the core to make it break. Double spiral fins of a uniform 240-mm pitch were welded onto the barrel, which was 2.2 m long with the upper 1 m being a chip reservoir. Some results from the test at T-3 were that (1) the drill usually stuck after proceeding 50 to 60 cm, requiring improvement of the chip-removing mechanism; (2) as long as the drill proceeded smoothly, the input current was less than 2 A, showing that the motor had enough power for chipping ice; (3) a drilling speed of up to 15 cm/min was obtained.

Concluding Remarks

Up to now, mostly thermal drills have been used in JARE ice-drilling projects because they are more easily made and more stable in operation than electrodrills. But it is evident that a thermal drill is far less effective than an electrodrill. While the drills 140, 140 Mk II and 160 have never reached a drilling speed of 1.5 m/hr, a primitive 400-W electrodrill easily reached a speed of 10 m/hr. This means that if the electrodrill could take the same length of core in one cycle as the thermal drills, it would shorten the overall drilling time of 450 m by more than 250 hours. Thus, the development of a reliable lightweight electrodrill is desirable, especially if an effective way can be found of removing ice chips. As for cutters, those of the CRREL mechanical drill seem successful as was recently shown by the performance of the drill made in Iceland (Árnason *et al.*, 1974).

Acknowledgments

Thanks are expressed to those who operated the drills in Antarctica and the Arctic, especially to Dr. H. Shimizu, Mr. T. Kimura, Mr. H. Narita and Mr. O. Watanabe, the leaders at Mizuho of JARE XI, XII, XIII and XV, respectively. Mr. T. Kimura designed the 1971 electrodrill. Special thanks are due to CRREL and ANARE for supplying the photographs and blueprints of the CRREL drill and the photographs of its Australian version, respectively, and also to Dr. K. Kizaki of Ryukyu University for obtaining them. Finally, thanks are due the Japan Polar Research Association for financial support to attend this symposium.

REFERENCES

Árnason, B., H. Björnsson and P. Theodórsson, 1974, Mechanical drill for deep coring in temperate ice: *Journal of Glaciology,* v. 13, no. 67, pp. 133-139.

Patenaude, R.W., E.W. Marshall and A. Gow, 1959, Deep core drilling in ice, Byrd Station, Antarctica: U.S. Army SIPRE Technical Report 60.

Shreve, R.L. and W.B. Kamb, 1964, Portable thermal core drill for temperate glaciers: *Journal of Glaciology,* v. 5, no. 37, pp. 113-117.

Ueda, H.T. and D.E. Garfield, 1968, Drilling through the Greenland Ice Sheet: U.S. Army CRREL Special Report 126.

Ueda, H.T. and D.E. Garfield, 1969a, Core drilling through the Antarctic Ice Sheet: U.S. Army CRREL Technical Report 231.

Ueda, H.T. and D.E. Garfield, 1969b, The USA CRREL drill for thermal coring in ice: *Journal of Glaciology,* v. 8, no. 53, pp. 311-314.

SOLID-NOSE AND CORING THERMAL DRILLS FOR TEMPERATE ICE

Philip L. Taylor
Department of Atmospheric Sciences
University of Washington
Seattle, Washington 98195

ABSTRACT

A generalized description is given of several hot-point drills, both coring and solid nose, which have been developed at the University of Washington for use in temperate ice. Drilling equipment is lightweight and readily transportable by small plane and hand-drawn toboggan, and has been used in support of research programs on Blue, Nisqually, and South Cascade Glaciers.

A coring hot-point drill is discussed which has been used to obtain continuous cores of firn and ice to depths of 90 m. The core is 15 cm in diameter, in lengths of 1.7 m, and its in situ *orientation is determined with a remote-reading inclinometer located in the drill barrel.*

Also discussed are solid-nose hot-points of several outside diameters which feature an inexpensive industrial cartridge heater which can be readily replaced in the field in case of burn-out. These drills range from 2.5 to 5 cm in diameter and advance at 6-8 m/hr utilizing 1000 to 2200 W. Depths of 210 m have been achieved. A tapered hot-point reamer has also been developed utilizing a 2200-W heater. This device has been used to control borehole refreezing, and as an aid in retrieving coring or drilling equipment which has become jammed in the hole.

Performance recommendations are given for solid-nose hot-points based on an optimized thermal efficiency.

Introduction

The development of portable, lightweight drilling equipment for glacier ice and firn has long been a challenge to glaciologists. For shallow depths (5 to 10 m) useful mechanical and steam drills have been described (Hodge, 1971; Kovacs *et al.,* 1973). For depths of 30-60 m, P. Kasser has designed a lightweight drill using pumped hot water (review by Shreve, 1962a). For greater depths the only practical solution seems to be offered by thermal drills utilizing electrical power. A review of the literature on thermal drills suggests that their development over the years has not evolved in any systematic way, but to have depended more on the needs of the particular drilling program and the materials and fabrication techniques that were at hand.

Successful solid-nose drills have been built with "Calrod" heater elements cast in copper (Stacey, 1960; Neave, 1968). Others have used hand-wound heaters (Shreve and Sharp, 1970).

A small-diameter drill using a silicon carbide resistance element has been described (LaChapelle, 1963). A small coring drill using a "Calrod" heater in circular form has been used (Shreve and Kamb, 1964).

High power densities are desired for fast and efficient drilling, yet lead to heater failures under the variable water and muck conditions encountered in a glacier borehole. Developments and experimentation will surely continue for a long time to come.

Large Coring Hot-Point Drill

A lightweight, portable thermal coring drill for temperate ice and firn has been developed and is illustrated in Figs. 1 and 2. The drill barrel is 6.75 in. (17 cm) in diameter by about 8.5 ft (2.6 m) long. The complete drilling system weighs approximately 400 pounds (180 kg) including a 3-kW gasoline-powered electrical generator. Transport by toboggan and set-up can be accomplished easily by two men. The barrel length was determined during the first field season to allow transport to the glacier aboard a small ski plane.

The drill is handled by suspending it from a collapsible 12 ft (3.7 m) tripod of aluminum tubing. A winch drum carrying 500 ft (150 m) of armored electrical cable is mounted between two legs, and the cable is run over a meter wheel sheave and down the hole. The winch is hand cranked for breaking off the core and for slowly approaching an end point. A variable-speed drill motor is used through a speed reduction to power the winch over long lifts. The meter wheel indicates depth of the drill in centimeters.

A fiberglass-epoxy pipe forms the body of the drill, supports all necessary components, and encloses the space for receiving the core. This pipe is lightweight and strong with a small wall thickness. It is electrically insulating, and its low thermal conductivity is useful in protecting the core.

Simplicity and ruggedness have been achieved in the drilling head by utilizing a ring of heavy nichrome wire operating at low voltage and high current (6 V, 360 A) directly exposed to the water. Special attention was given to the mechanical support and to the electrical connections to the heater ring to achieve a uniform temperature distribution around the ring, eliminating the cold spots and increasing the thermal efficiency. Power is fed to the heater ring through two sheet-copper conductors mounted on the outside of the fiberglass-epoxy pipe. These sheet conductors are connected to the secondary winding of a step-down transformer rated at 2.2 kW mounted in the upper section of the drill barrel above the core space. The transformer primary receives 220 V, 60 Hz single-phase power at about 10 A through the supporting cable. The transformer and all the connections to it are designed for direct water immersion.

Interchangeable thermal drilling heads are used for wet, impermeable ice and for dry drilling in firn. The heads are readily exchanged, are electrically identical, but differ in mechanical construction since the firn head operates at a higher temperature and has negligible side melting due to the lack of water. When drilling in firn, power input is reduced to about 70 per cent of full to prevent damage to the heater ring when part or all of it becomes suspended in air. As soon as the water table is reached the more efficient ice head is installed, and full power is applied.

A short loop of heavy resistance wire located on the inside diameter of the barrel near the drill head melts a small longitudinal groove in the core as the drill advances. This groove defines

Figure 1. The thermal coring drill. Winch installation on the near side of the supporting tripod.

Figure 2. Core removal and display.

169

the orientation of the core in the barrel even though the core may fracture or move during retrieval. As the core fills the space available in the barrel a spring-loaded push rod activates a switch which signals the operator through a lamp circuit in the control box. Drill controls are illustrated in Fig. 3.

The core is broken off and captured by spring-loaded sharp steel fingers near the drill head. Pulling on the drill forces the fingers into the core. At a force of about 300 pounds (140 kg) in ice to about 500 pounds (230 kg) in firn the core breaks across and is held in the barrel. The core is removed on the surface by holding these fingers in a retracted position with lock pins.

The top end of the drill barrel consists of a tapered aluminum cone containing a heater which may be activated from the surface in case the drill becomes stuck in the borehole during retrieval. This "back-out" heater remains untested as no difficulty has ever occurred.

A remote-reading two-axis inclinometer and magnetic compass utilizing separate conductors in the cable is located in the upper section of the drill barrel. This device enables the operator to determine the *in situ* orientation of the drill barrel after drilling the core and before breakoff. Structure noted in the core after retrieval can be referenced to the longitudinal groove of the core marker and to finger scratches, and hence to the drill barrel, so that the *in situ* orientation of the core immediately prior to breakoff can be determined. Experience in the field has been that the core can be orientated to within 5-8 degrees in azimuth and to about 0.25-0.5 degree in tilt. Typical core cross sections are illustrated in Fig. 4 (90-m hole, Blue Glacier, 1971). The cores are used for stratigraphic studies of firn in the accumulation areas, for structural measurements of ice fabrics, grain size and shape, and for microscopic examination of vein structure and bubble properties.

Figure 3. Coring drill controls. Operator is reading the inclinometer prior to core breakoff and retrieval. Drill power controls in the foreground.

170

Drill rate in temperate ice at 210-220 V and 8.5 A (1.8-1.9 kW) at the generator end of the drill cable averaged about 1.7 m/hr, with a range of 1.5 to 2.0 m/hr (40- and 90-m holes, Blue Glacier, 1971). Voltage drop along the cable was estimated to be about 23 V, indicating a power at the drill transformer of 1.6-1.7 kW. Tests in block ice in the lab have given rates of 2.5 m/hr with 208 V on the transformer. At the full rated power of 220 V and 10 A on the transformer, a drill rate of about 2.8 m/hr would be expected. At the required reduced voltage input at the generator of 180 V for snow and firn the drill rate varies from 1.5 to 2.5 m/hr, depending on density and ice layers. Core diameter is about 6 in. (15.2 cm) and the hole is about 7 in. (17.8 cm) in diameter. Core length is usually 1.65-1.7 m.

Figure 4. Samples of a continuous vertical core taken near the base of the icefall, Blue Glacier, Washington, 1971. Numbers indicate depth from the surface in meters; scale divisions are decimeters.

171

Hot-Points

The solid-nose hot-points that we have developed are modeled after the highly successful Shreve unit (Shreve and Sharp, 1970), the significant difference being in the utilization of an inexpensive (about $7) industrial cartridge heater. The hot-points are opened easily for maintenance and for replacement of the heater in case of burn-out. A cross-sectional view of the 2 in. (5.08 cm) diameter hot-point is shown in Fig. 5. The heater is rated at 2200 W, 220 V, and is inserted into the copper slug with a light push. The contact surface between the heater and copper is treated with a commercial silver plating compound to improve and maintain the heat transfer. This increases the reliability of the heater by reducing its operating temperature. The heater is wired through the watertight plug and is easily removed as a unit with a small pulling tool as illustrated in Fig. 6.

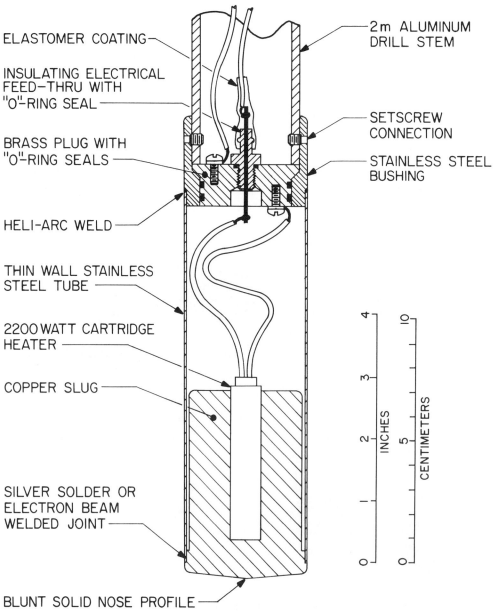

ELASTOMER COATING

INSULATING ELECTRICAL FEED-THRU WITH "O"-RING SEAL

BRASS PLUG WITH "O"-RING SEALS

HELI-ARC WELD

THIN WALL STAINLESS STEEL TUBE

2200 WATT CARTRIDGE HEATER

COPPER SLUG

SILVER SOLDER OR ELECTRON BEAM WELDED JOINT

BLUNT SOLID NOSE PROFILE

2m ALUMINUM DRILL STEM

SETSCREW CONNECTION

STAINLESS STEEL BUSHING

INCHES

CENTIMETERS

Figure 5. Cross section of the solid-nose hot-point drill.

To maintain thermal efficiency the nose is blunt (Shreve, 1962b; Shreve and Sharp, 1970) and a thin-wall stainless-steel tube of low thermal conductivity was chosen for the body. The hot-point is guided by a 2-m-long drill stem of aluminum tubing weighted with about 15 pounds (7 kg) of lead. This weight improves the thermal efficiency by decreasing the thickness of the water layer at the nose of the drill.

The surface platform is illustrated in Fig. 7 and contains a cable supply drum, meter wheel sheave, and power control box. The drum contains 500 ft (150 m) of cable, an electrical slip-ring, and friction drag.

Performance checks over several field seasons and hundreds of meters of boreholes have given typical drilling speeds of 5-6 m/hr at about 1300 W power input. Tests in block ice with 1800 W input gave a drill speed of 7.5 m/hr. Calculated thermal efficiencies range from 0.75 to 0.85, indicating that if generator and cable limitations could be overcome the drilling speed at the rated input power of 2.2 kW would be approximately 9 m/hr. These drills have been used for the installation of devices for measuring vertical strain rates in glaciers (Rogers and LaChapelle, 1974) and for probing glacier water systems (S.M. Hodge, South Cascade Glacier, personal communication, 1974).

The use of cartridge heaters allows ease of design in scaling to different hot-point diameters, and for ease of field substitution for different wattages and voltages depending on the electrical generator that is available. Hot-points of similar design with diameters of 1 in. (2.54 cm) and 1.25 in. (3.18 cm) have also been constructed using a similar cartridge heater rated at 1000 W, 110 V or 220 V, and measuring 0.5 in. (1.27 cm) diameter by 1.5 in. (3.8 cm) long. The drill cable termination is standardized to allow interchange of the different size hot-points and the reamer described below.

Figure 6. Solid-nose hot-point assembled, and with plug and heater assembly removed with small pulling tool. Scale in inches.

Figure 7. Solid-nose drill platform. Operator holds the drill stem; cable has been lifted off the meter wheel sheave. Cable drum and power control box on the right.

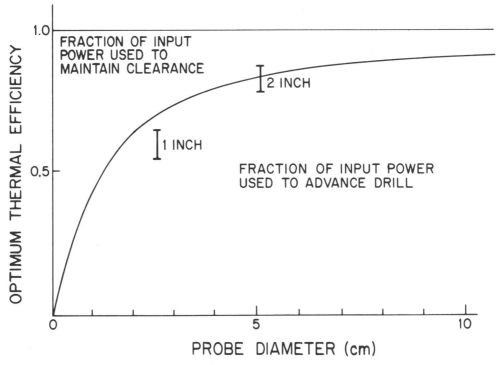

Figure 8. Optimum thermal efficiency for solid-nose hot-point drills operating in wet temperate ice. Clearance diameter is 5 mm between probe and hole. Bars show range of observed values.

Performance of the 1 in. (2.54 cm) hot-point is typically 6-8 m/hr at power inputs of 450-700 W, indicating that thermal efficiencies are 0.55 to 0.65. At least 1800 m of hole have been drilled, the maximum depth being 210 m. The boreholes have been used for verifying ice thickness (Hodge, 1974), planting thermistors (Harrison, 1975) and for probing glacier water systems (S.M. Hodge, South Cascade Glacier, personal communication, 1973).

Power must be reduced to about 25 per cent of maximum when drilling in snow to prevent burn-out. A long aluminum coring tube is used to get through most of the snow cover. A portable steam drill has also proven valuable (Hodge, 1971).

Another cartridge heater 0.5 in. (1.27 cm) diameter by 20 in. (51 cm) in length, rated at 2200 W at 220 V has been utilized in the development of a tapered borehole reamer measuring about 2.5 in. (6.3 cm) in diameter and 24 in. (61 cm) long. The reamer is machined from a solid aluminum round bar, with the heater inserted in a long hole on the cylinder axis. The electrical and mechanical connections are compatible with the solid-nose drill-cable termination. The reamer has been successful in controlling refreezing in existing boreholes to ensure safe passage of instrumentation. In the event that equipment does get jammed in the hole, or a cable tangle occurs, the reamer has proven to be indispensable in working its way to and around the obstruction and enlarging the cavity to enable retrieval.

Comments on Thermal Efficiency in Hot-Point Design

The solid-nose hot-point design problem is an interesting one in that for a given hole size and power available there is a whole family of drills of acceptable design depending on what trade-offs have been made between thermal efficiency, probability of irreparable damage or loss, and the cost of construction. Thermal efficiency of a drill for wet, temperate ice is considered here to be based on the probe cross-sectional area, the drilling speed, and the power input to the hot-point assembly. Assuming that all the input power melts ice, the thermal efficiency is also equal to the ratio of the area of the probe to the area of the hole. Beyond some point it does more harm than good to increase the thermal efficiency as the drill is more apt to hang up in the hole and be lost. A drill of low thermal efficiency wastes fuel and time as it makes a bumpy hole of irregular cross section because of the larger percentage of energy which goes to heat water which moves up and sideways and melts the ice in an uncontrolled manner.

The probability of a drill getting stuck in the borehole for a given stem length seems to be largely a function of the clearance between the drill and the hole. An intuitive guess based on observation and experience suggests a difference of about 0.2 in. (5 mm) between the probe diameter and the hole diameter as a good design guideline to achieve a useful, efficiently drilled hole of smooth and constant cross section with a minimum of risk. This means that there is an optimum thermal efficiency which is a function of the drill diameter, with smaller drills operating at lower thermal efficiency. This is illustrated in Fig. 8.

Fortunately, the tendency in scaling hot-points is in this direction, since power input for constant drill rate is proportional to probe area, while side losses tend to be proportional to probe circumference. One could then expect the ratio of side loss to power input to vary somewhat as the inverse of the probe diameter. This is close to the desired effect, and greatly simplifies the design process by making the optimum efficiency easier to achieve.

Figure 9 illustrates how drill rate and power input vary for drills operating at this optimum thermal efficiency, and is presented here as a design guide. The borehole diameter in all cases will be 0.2 in. (5 mm) greater than the probe diameter.

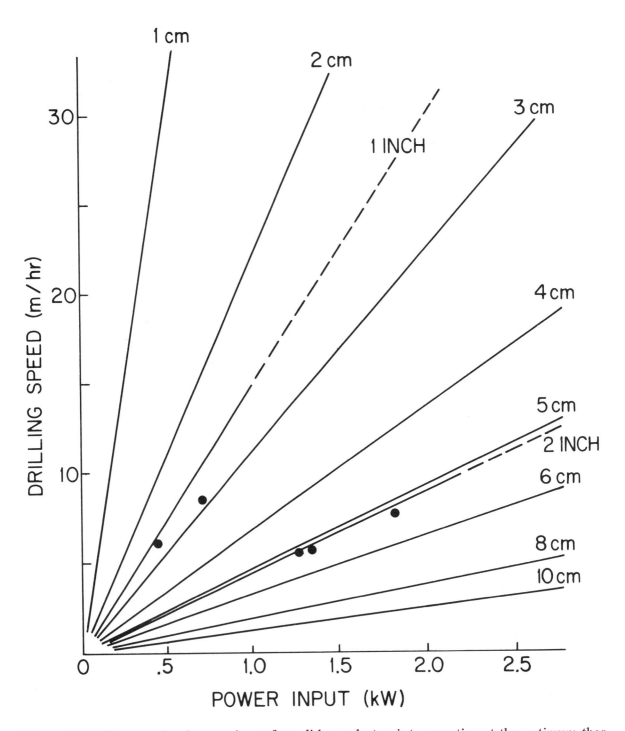

Figure 9. Drilling speed and power input for solid-nose hot-points operating at the optimum thermal efficiency shown in Fig. 8. Observed values for the 1-in. and 2-in. drills are shown as dots; dashed extensions are for power input above the cartridge rating.

Acknowledgments

These developments were supported by NSF Grants GA 1516, GA 28544 and GU 2655. Thanks are given to E.R. LaChapelle, C.F. Raymond, S.M. Hodge, and W.D. Harrison for their many helpful suggestions for design improvements, the use of their photographs, and the opportunity to participate in their field research programs.

REFERENCES

Harrison, W.D., 1975, Temperature measurements in a temperate glacier: *Journal of Glaciology,* v. 14, no. 70, pp. 23-30.

Hodge, S.M., 1971, A new version of a steam-operated ice drill: *Journal of Glaciology,* v. 10, no. 60, pp. 387-393.

Hodge, S.M., 1974, Variations in the sliding of a temperate glacier: *Journal of Glaciology,* v. 13, no. 69, pp. 349-369.

Kovacs, A., M. Mellor and P.V. Sellmann, 1973, Drilling experiments in ice: U.S. Army CRREL Technical Note (unpublished).

LaChapelle, E., 1963, A simple thermal ice drill: *Journal of Glaciology,* v. 4, no. 35, pp. 637-642.

Neave, K.G., 1968, Glacier seismology: Masters Thesis, Department of Physics, University of Toronto, Toronto, Canada, pp. 23-26.

Rogers, J.C. and E.R. LaChapelle, 1974, The measurement of vertical strain in glacier bore holes: *Journal of Glaciology,* v. 13, no. 68, pp. 315-319.

Shreve, R.L., 1962a, Review: The thermal ice drill of P. Kasser: *Journal of Glaciology,* v. 4, no. 32, pp. 234-235.

Shreve, R.L., 1962b, Theory of performance of isothermal solid-nose hot points boring in temperate ice: *Journal of Glaciology,* v. 4, no. 32, pp. 151-160.

Shreve, R.L., and W.B. Kamb, 1964, Portable thermal core drill for temperate glaciers: *Journal of Glaciology,* v. 5, no. 37, pp. 113-117.

Shreve, R.L., and R.P. Sharp, 1970, Internal deformation and thermal anomalies in lower Blue Glacier, Mount Olympus, Washington, U.S.A.: *Journal of Glaciology,* v. 9, no. 55, pp. 65-86.

Stacey, J.S., 1960, A prototype hot-point for thermal boring on the Athabaska Glacier: *Journal of Glaciology,* v. 3, no. 28, pp. 783-786.

THERMAL AND MECHANICAL DRILLING IN

TEMPERATE ICE IN ICELANDIC GLACIERS

Páll Theodórsson
Science Institute
University of Iceland
Reykjavik, Iceland

ABSTRACT

A thermal ice corer was constructed in Iceland in 1968 and used for drilling a 108-m-deep hole in 1969. For shallow holes an electrical drive was added to a SIPRE coring auger in 1969 and this has been used for making a number of 10- to 20-m-deep holes.

A larger mechanical corer was then constructed for deep holes and used to drill a 415-m hole into the ice cap of Vatnajökull. The experience gained in this work and some possible refinements in the drilling technique are discussed.

Introduction

The hydrology of Iceland has been studied extensively at the Science Institute of the University of Iceland using deuterium and tritium as natural tracers. Glaciers are an important part of this study as they cover about one-tenth of the area of the country.

Initially the isotopic investigations called for shallow holes that were drilled with a SIPRE coring auger, but in order to better understand isotopic processes in temperate glaciers (Árnason, 1970) deeper holes were needed. This was solved by making a thermal corer with which two holes were drilled in 1968, an initial hole to a depth of 30 m and a second hole in the ice cap of Bárdarbunga in Vatnajökull glacier to a depth of 42 m, where the corer got stuck and could not be recovered.

The next summer a more serious attempt was made to drill a deep hole into Bárdarbunga with a new thermal corer of almost identical construction. It reached a depth of 108 m. This corer worked very well in firn but when solid ice was reached at a depth of about 30 m the drilling speed fell abruptly and the diameter of the ice core became smaller. No core was recovered below a depth of 102 m, probably because of excessive melting caused by a thick ash layer.

From this experience it was considered unlikely that a thermal corer could be used successfully for deep drilling in Icelandic glaciers.

Using a thermal corer in firn the meltwater will possibly disturb the isotopic ratio of the ice. In order to get undisturbed cores from the upper layer of the glaciers a SIPRE coring auger was used. However, it is difficult or impossible to drill holes deeper than about 5 m in temperate glaciers with the hand-driven SIPRE coring auger as one must maintain fast and even rotation of the coring auger. An electrical drive was therefore attached to the corer and with this modification it proved quite efficient for drilling to a depth of about 20 m.

Initially our interest was only for the surface layers of the glaciers but when it had been demonstrated by the combined pioneering work of Dansgaard *et al.* (1969) at the University of Copenhagen and by Langway and Hansen (1970) at CRREL that polar glaciers preserve a record of past climatic fluctuations, interest increased for deep drilling into the glaciers of Iceland. It was considered possible that the ice cap of Bárdarbunga in Vatnajökull might preserve a similar record of climatic fluctuations in Iceland since its settlement some 11 centuries ago. Such a record would be of great interest. It could be compared to considerable information in written records for the study of the reliability of the isotopic data and eventually used to extend and implement the written record.

Finally the ice core would be of interest for various other studies: the measurement of tritium and various trace elements, the study of volcanic ash layers and the physical characteristics of the ice.

A mechanical ice corer was therefore designed and constructed with financial support from the International Atomic Energy Agency in Vienna and used to drill a 415-m-deep hole into Bárdarbunga in 1972. Because of failure in the cable the bottom of the glacier could not be reached.

This paper describes the experience gained in this work and discusses some future possibilities. It is the result of work of a team at the Science Institute of the University of Iceland as well as others. The skillful work of Karl Benjaminsson, who made most of the mechanical equipment and made an important contribution to the design, is gratefully acknowledged.

Thermal Ice Corer

A simple thermal corer described by Shreve and Kamb (1964) was generally followed. Figure 1 shows a cross section of the thermal corer. The heating element is made from a 0.6-mm-diameter heating wire that is wound tightly into a helix and then pressed into a 6-mm-wide circular groove in the heater housing with thin mica insulation between the housing and the wire. The free space in the groove is then filled with alundum cement. One end of the wire is soldered to the housing but the other end is fed through a teflon insulator in the upper annular plate of the heater housing. The plate is soft-soldered to the lower part of the heater housing. An annular thermal insulator of teflon is inserted between the heater housing and the rest of the thermal corer. The total length of the corer is 120 cm.

In the first corer a polyethylene tube was inserted inside the core tube for thermal insulation, but on two subsequent thermal corers the tube was omitted, making no apparent difference.

About 250-350 W were usually applied to the heater, this being the power that the generator could produce. The heater could withstand more power but with 600 W the useful lifetime of the heating element became too short.

180

This thermal corer proved efficient for drilling through firn. The drilling speed was about 2 m/hr and the core radius was only about 2 mm less than that of the heater and the radius of the hole about 2 mm larger than that of the heater. The corer was used once in Sweden to drill through firn but the drilling speed was much less than 2 m/hr as the heater melted more of the core and the diameter of the hole was larger.

When the corer reached solid ice at a depth of 30-40 m the drilling speed fell abruptly to less than a meter per hour and it varied considerably from one run to the next. The core showed clearly the reason: instead of recovering a core with a diameter of 45 mm, as was being done in the firn, the core usually had a diameter of 35 mm or less. Sometimes no core was recovered at all and the core recovery averaged less than 50 per cent.

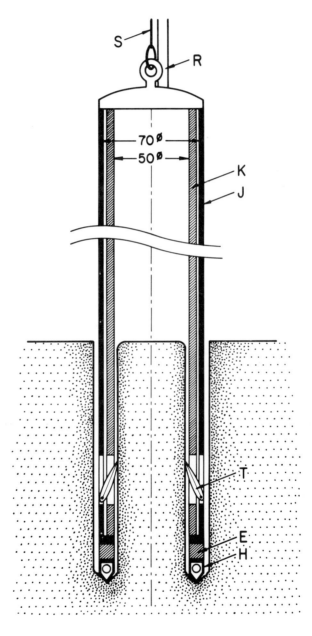

Figure 1. Cross section of thermal drill. H: heater. E: thermal insulator of teflon. T: pawl for breaking ice core. J: core barrel. K: insulating tube of polyethylene. R: cable. S: steel wire.

During a few runs in solid ice the drilling speed was quite high compared to an average run and a good core would be recovered with a diameter of 35-40 mm along its full length of 1 m. In the next run the corer would penetrate faster than normal in solid ice and produce a reasonably good core. During the drilling we were aware that there was water in the hole but we had no means to measure the depth to the water table. Some crude estimates were made by fastening pieces of tissue paper to the cable at varying heights above the corer. From a few such unsystematic observations we got the impression that the water table was usually some meters above the corer and it seemed lower when good cores were being recovered. This is, however, not in accordance with the experience of the deep drilling three years later, made about 5 km away at an elevation some 200 m lower, where the water table was very stable (discussed below).

It seemed clear that the meltwater trapped in the solid ice gave rise to convection currents at the bottom of the hole, causing excessive melting and a slow drilling rate. Attempts were made to hinder such convection currents by fastening rubber blades on the top of the thermal insulator in an effort to isolate the bottom from the rest of the hole. This seemed to have no positive effect.

When a depth of 102 m was reached the drilling speed became still slower and no core was recovered at all. Deep drilling three years later showed that a thick ash layer can be expected at this depth.

After this experience we were not optimistic about the use of a thermal ice corer for deep drilling in glaciers in Iceland. Experience in other countries indicated, however, that a thermal corer with a direct heater might yield better results. Such a corer was made together with the mechanical ice corer for drilling in Bárdarbunga in 1972. The heater consisted of a single circular turn of bare wire with a diameter of 3 mm. The diameter of the circle was 115 mm. Two short wires were soldered vertically to the circular heater and this served both for fastening the heater and as electrical leads for the low voltage supply. A 3-kW transformer was fastened to the upper end of the core barrel. This corer was tried in solid ice at Bárdarbunga but the drilling speed was very slow and this attempt was soon abandoned, along with any prospect of using a thermal corer for deep drilling in glaciers in Iceland.

Electrically Driven SIPRE Coring Auger

The SIPRE coring auger has been modified by adding an electrical drive to it. A 750-W electrical hand drill is fastened to the top rod for rotating the ice corer. The speed of rotation is 250 rpm, but this proved too fast and the speed was decreased by frequently interrupting power to the motor. About 100-150 rpm would probably be optimal.

New rods were made from thin-walled steel pipes with special connecting pieces soldered to their ends. These pieces (Fig. 2) have a rectangular cross section and the lower one fits into the upper one; the latter rests on a rim at the end of the circular pipe. A wing screw holds the two connecting pieces together and, because of the rectangular cross section, there is no strain on the screws during drilling The rods are different lengths (1, 2 and 3 m) because it is most convenient to add only 1 m to the total length of the rod string each time.

When this arrangement was first tried there was no means to center the rods in the hole and they had a tendency to oscillate and beat against the wall. To remedy this, centering pieces (Fig. 2) were put on the rods 2-3 m apart, nearly eliminating the oscillation. In this form the ice corer has been used extensively to drill 5- to 15-m-deep holes.

In glaciological work there is a frequent need to drill holes of 30 m or less and this should preferably be made with as simple equipment as possible. With further modifications the SIPRE coring auger would be suitable for this kind of work.

With the present technique it is easy to drill to 10 m but below this it becomes increasingly difficult, mainly because of the weight of the rods. It should, however, not be difficult to add a simple winch to the equipment using the same motor as for the corer.

Some kind of platform should be added, both for centering the rods at the upper end as well as for locking the rod string while adding a new rod or taking one off.

One of the most serious drawbacks of the SIPRE coring auger is that it takes only a 30- to 40-cm-long ice core in each run. Attempts to increase this have been futile. With the faster and more even rotation of the electrical drive it should be possible to lengthen the corer and decrease the width of cutting. The ice chips would then take less space and this would make room for a longer core.

The Mechanical Corer for Deep Drilling

The equipment used for drilling a 415-m-deep hole into the ice cap of Bárdarbunga in Vatnajökull glacier has already been described and some discussion of the experience using it presented (Árnason *et al.*, 1974). Some features of the drill and corer will be described in more detail here and some design considerations discussed as well as the experience gained. Some suggested improvements of the drilling equipment and corer will also be discussed.

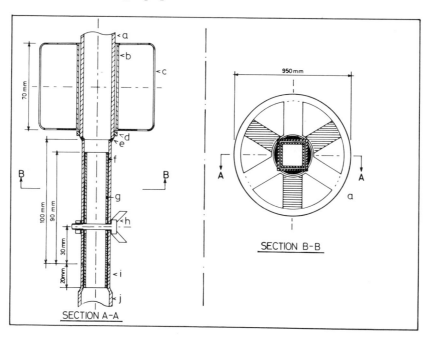

Figure 2. Connecting pieces of drill rods and centering pieces. a: circular rods. b: centering pieces. d: rim on which centering piece rests. e: soft solder. f: upper rectangular connecting piece. g: lower connecting piece. h: wing screw.

When the design of the corer began many uncertainties confronted us as we were guided by very limited practical experience. Our use of the electrically driven SIPRE coring auger gave us valuable hints and the experience the author gained in participating in an American-Danish drilling project in Greenland in 1971 was of much help.

It was of great importance to know whether there would be water in the hole because it would presumably strongly influence the transfer of the ice chips from the cutting bits. Further, if the hole would be partially filled with water this would counteract the closure of the hole that was considered to be one of the most serious obstacles to deep drilling in temperate ice.

Experience from the 108-m hole drilled with the thermal corer indicated that there would be water at the bottom of the hole most of the time after solid ice had been reached. It was therefore assumed that the drill and corer would be submerged in water after the firn had been penetrated. Some difficulties could be expected in the firn with the mechanical corer and this was partly the reason for also making a thermal corer, as experience had shown that it worked very well in firn.

Figure 3 shows a cross section of the drill and corer. Briefly, it operates as follows. A 2.0-m-long core barrel tube rotates inside a stationary outer tube. Two cutting bits on the lower end of the core barrel cut a 15-mm-wide groove into the bottom of the hole. The ice chips are carried up into the storage compartment above the core barrel by a helical rib on the barrel. After each run the storage compartment is emptied.

The total length of the ice corer is 6 m. In the *cable termination cylinder* on the top of the corer there is a strong helical spring that carries the weight of the drill and corer. A device senses the compression of the spring, indicating when the corer is resting on the bottom of the hole with more than a certain fraction of its total weight. A pin touches an insulated spring when about 70 per cent of the weight is resting on the bottom of the hole. An ohmmeter measures the resistance between the spring and the pin. Initially this was meant to be an ON-OFF device only but as the gap is filled with water when the drill and corer are in the hole the resistance begins to fall when the pin approaches the spring, indicating that the drill and corer are resting with increasing weight on the bottom of the hole. This proved to be of much additional help during drilling and has shown us that a potentiometer should replace the spring in future work to give a continuous indication of the weight on the bottom of the hole.

The *electric motor* was a 3-phase 380-V AC, 3-hp motor, but it failed and was replaced by a 3-phase 220-V AC, 2-hp motor. Both motors are of the same type and are made for submergible pumps.

A *planetary gear* taken from a starter in a DC-3 airplane (omitting one stage) was used to reduce the speed of rotation to 150 rpm. Later experience indicated that this speed was suitable. The outer diameter of the gear was the main factor in deciding the diameter of the corer.

Below the gear there are three torque skates that are pressed against the wall to prevent the rotation of all parts but the core barrel. The length of the torque skates is 15 cm. In solid ice these gave sufficient force to keep the outer tube from rotating. In firn the torque seemed to be sufficient also but the cable with its large diameter (22 mm) may have helped provide some of the torque needed. While drilling through the firn we were occupied with various minor problems so that close attention was not given to the performance of the torque shoes.

Between the torque skates and the core barrel there is a 2-m-long annular space for storing the ice chips during the drilling, the *storage compartment*. The storage compartment is emptied after each run through a hatch on the cover tube.

The *core barrel* (inner/outer diameters 91/95 mm) is 2.0 m long and the space between it and the cover tube (inner/outer diameters 106/110 mm) is only 6 mm. The ice chips are transferred during drilling from the cutting bits to the storage compartment by a pair of auger flights on the core barrel. These are made of rectangular steel bars (5x5 mm) silver-soldered to the barrel. The pitch of the flights is 20 cm at the lower end and increases to 26 cm at the upper end.

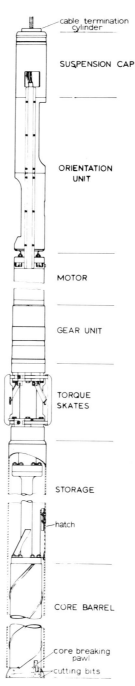

Figure 3. Cross section of the mechanical drill.

Considerable attention was given to the removal of the ice chips. The possibility of letting them float freely in the hole, which we assumed would be partially filled with water, was considered. We feared, however, that the corer might cut into a crevasse in the ice, the water would run out of the hole, and the ice chips might then clog the lower part of the hole. This idea was therefore abandoned.

The possibility of partially melting the ice chips was then considered. Water was to be circulated with the help of a pump above the core barrel and heater elements were to be inserted into the storage compartment, which was considerably smaller than in the final design. It was hoped that the stream of water would help in moving the ice chips up between the core barrel and outer tube and then divert the ice chips to the heater elements. This arrangement was also considered useful in case the corer would get stuck in the hole, as the circulation of the heated water would help in freeing the corer. Laboratory tests with a prototype indicated, however, that the water circulation helped little in moving the ice chips up into the storage compartment, so this design was therefore abandoned.

After this the corer was designed with a storage compartment large enough to retain all the ice chips from a single run and the heaters were only considered as an additional aid that might be resorted to in case the corer showed a tendency to get stuck. The length of the storage compartment proved to match well the length of the core barrel, as the ice chips were tightly packed there in the few runs when an ice core of full length (2.0 m) was drilled.

Initially there were no auger flights in the storage compartment. After having drilled in solid ice for some time the ice chips seemed to be pressed rather firmly together at the bottom of the storage compartment. A 15-cm-high (approx.) double auger flight was therefore welded to the axle just above the core barrel. No further difficulties of this type were observed.

When the corer was being designed it was first considered helpful to have the pitch of the auger flights greater at the lower end so that the ice chips would be carried more firmly away from the cutting bits. A test with a prototype of the corer showed, however, that the reverse was the case; i.e. the ice chips became clogged when they reached the point where the pitch decreased. When the pitch was changed, increasing slowly all the way up along the core barrel, this difficulty disappeared.

Two *cutting bits* are fastened with screws to the core barrel as well as a pair of core catchers for breaking the core at the end of each run.

Although many difficulties were met in the drilling work and the bottom of the ice cap was not reached, an important objective was achieved: the corer worked successfully in the last phase of the work and a total depth of 415 m was reached. The estimated thickness of the glacier at the site of drilling is 450-500 m.

Three serious difficulties were met: (1) weakness of the cable, (2) low drilling rate in solid ice, and (3) frequent sticking of the corer at the bottom of the hole at the end of a drilling run.

Creeping occurred in the braided steel armor of the cable under load resulting in the load being transferred to the copper leads which were occasionally stressed beyond the breaking point. A final break in a power lead stopped the drilling temporarily at a depth of 298 m. A new cable was borrowed from CRREL and the drilling was resumed after a delay of three weeks. The length of the cable limited the final depth of the hole to 415 m. The diameter of the hole was not measured but judging from the smooth movement of the corer no closure of the hole could be sensed. The hole was, however, filled with water below a depth of 32 m, just below the limit between firn and ice. The water level was quite stable. We once used the outer tube of the corer, after the core barrel had been removed and the lower end of the outer tube closed, to bail water out of the hole. The amount of water bailed out was equivalent to a water column of about 7 m but the water level in the hole changed less than 1 cm.

The most serious difficulty in the drilling work was the low drilling rate. A test with a proto-type in the laboratory had indicated that the corer would cut through 1 m in about 5 minutes. When we were drilling through the firn, where we had anticipated some difficulties in transport-ing the dry ice chips up into the storage compartment, the drilling speed was satisfactory. When we reached solid ice just below 30 m the drilling speed was reduced considerably. Instead of drill-ing 2 m in 5-10 minutes a penetration of 40-80 cm would take 20-50 minutes and then the corer would not go deeper. Shaking the cable and turning the corer motor on and off helped consider-ably but to a lesser degree as the hole became deeper. As long as the drilling speed was high enough to make it likely that we could drill through the estimated 450-500 m of the ice cap, the main emphasis was on continuing the drilling.

Gradually evidence accumulated indicating that the ice chips were freezing together on the cutting bits, thus reducing and eventually stopping the drilling. This difficulty was successfully overcome by introducing an antifreeze mixture to the bottom of the water-filled hole. A poly-ethylene bag with about 180 ml of isopropyl alcohol was tied to the end of the core barrel. When the motor of the corer was started, the bag burst and the alcohol mixed with the water by the rotating core barrel. This lowered the freezing point at the bottom of the hole and hindered the freezing of ice chips around the cutting bits. The third major problem was the frequent sticking of the corer at the bottom of the hole at the end of a drilling run. After we started using the anti-freeze mixture it would take only about 5 minutes to drill 100-140 cm. The drilling speed would then fall abruptly, presumably because the alcohol had been diluted so much that the ice chips started freezing around the cutting bits. The motor was then stopped and the corer raised 20-30 cm by turning the winch by hand. The corer would then sometimes lodge fast in the hole. In such cases a tension of about 600 kg was applied to the cable. In most cases this would be sufficient to free the corer but sometimes it would not move. We then waited until it got loose by itself under this tension. This could take up to an hour but twice we had to wait longer. Once we waited for some eight hours and the corer did not get loose until we had increased the tension by about 20 per cent.

This sticking is probably caused by the ice chips freezing together at the bottom of the hole outside the outer tube. This could probably be avoided if the drilling stopped before the alcohol became diluted so that the ice chips begin to freeze together. However, because a major part of the total drilling time was taken up by raising and lowering the drill, it was tempting to take as much core in each run as possible. Further, as the day shift was competing secretly against the night shift little time was left for experimenting during this final stage of the drilling. Therefore we lived with the problem rather than solving it.

Having discussed the major problems of the drilling some general remarks follow about the performance of the corer. The coring motor seemed to be working lightly most of the time. There was an ammeter on all 3 phases of the motor in the corer. Hanging freely in the hole the motor would take about 4.0 A. Under light load during drilling the current would rise to 4.3 A, occasionally to 4.5-5.0 A, and higher in extreme cases. It should be remembered that the load will not be proportional to the current because of varying phase difference between the voltage and current. It would have been more useful to have a wattmeter measuring the load of the motor. The load seemed to depend strongly on the weight with which the corer was resting on the bottom of the hole. The warning device, that was to show when this exceeded 70 per cent of the total weight of the corer, was inoperative most of the time because of a break in the signal leads of the cable. Most of the drilling was therefore done under unfavorable conditions. As already noted the corer got stuck at a depth of 17 m and once the motor burned out because of excessive load, presumably because it was continuously resting on the bottom of the hole with its full weight. The last 118 m were drilled with a cable borrowed from CRREL because of a break in the power leads in the initial cable. During this time the load warning device was operative and this helped much in giving smooth drilling.

Drilling Facilities and Suggestions

Good working conditions are important for prolonged drilling on glaciers. How this is solved will depend on the conditions. In our case we provided excellent working conditions by digging a large pit (7.0 x 2.7 m wide and 4.2 m deep) and covering it with a roof of polyethylene foil fastened to a wooden frame. Drift snow collected on the roof, requiring considerable work in keeping the roof free. A combination of pit and surface shelter would probably have been better.

Insufficient attention was given to the shelter for the power plant. It should have good cooling during calm weather but should be sufficiently well closed during storms so that drift snow will not make it wet. Care should be taken that the exhaust is vented so that it will not melt the surrounding snow. Finally it is useful to have some arrangement in the power plant to dry clothing. As drift snow is likely to collect around the shelter of the power plant, the plant should be on a sledge so that it can be moved easily and the cables kept free from being buried in the drift snow.

For future drilling operations, if a hole of 400 m or less is being drilled, the core barrel and storage compartment should be shortened to about 120 cm each instead of the present 200 cm.

The same method of delivering the antifreeze mixture to the bottom of the hole might be used. It would, however, be desirable to pump the alcohol continuously to the bottom of the hole during drilling, thereby probably eliminating the difficulty we experienced in freeing the corer at the end of the run. For some studies it is undesirable to add any foreign fluid to the hole. A heater on the lower end of the outer tube might solve the problem of the freezing ice chips instead of using the alcohol.

The time it takes to remove the core and empty the ice chips from the storage compartment should be shortened. It should be possible to reduce this from the present 15 minutes to 5 to 7 minutes.

The most probable of the more serious problems one can expect in deep ice-core drilling is getting the corer stuck in the hole. In such cases it is important to know reasonably well how much tension can be applied without impairing the cable and winch and have reasonably good indications of the tension in the cable. We solved this by pulling the winch handle with a spring balance. A more sophisticated arrangement would be desirable, but a spring balance should be available.

REFERENCES

Árnason, B., 1970, Exchange of deuterium between ice and water in glaciological studies in Iceland: *Isotope Hydrology 1970.* Proceedings of a symposium on use of isotopes in hydrology held by the International Atomic Energy Agency in Vienna, 9-13 March 1970, pp. 59-71.

Árnason, B., H. Björnsson and P. Theodórsson, 1974, Mechanical drill for deep coring in temperate ice: *Journal of Glaciology,* v. 13, no. 67, pp. 133-139.

Dansgaard, W., S.J. Johnsen, J. Møller and C.C. Langway, Jr., 1969, One thousand centuries of climatic record from Camp Century on the Greenland ice sheet: *Science,* v. 166, no. 3903, pp. 377-381.

Langway, C.C. Jr. and B.L. Hansen, 1970, Drilling through the ice cap: Probing climate for a thousand centuries: *Bulletin of the Atomic Scientists,* v. 26, no. 10, pp. 62-66.

Shreve, R.L. and W.B. Kamb, 1964, Portable thermal core drill for temperate glaciers: *Journal of Glaciology,* v. 5, no. 37, pp. 113-117.